Number Theory

Number Theory

Sudarsan Nanda

Professor of Eminence and Research Chair

KIIT University, Bhubaneswar 751024, India

ALLIED PUBLISHERS PRIVATE LIMITED

New Delhi • Mumbai • Kolkata • Lucknow • Chennai • Pune

Nagpur • Bangalore • Hyderabad • Ahmedabad

ALLIED PUBLISHERS PRIVATE LIMITED

Regd. Off. : 15 J.N. Heredia Marg, Ballard Estate, Mumbai-–400001, Ph.: 022-22626476
E-mail: mumbai.books@alliedpublishers.com
Amber Chambers, 28-A Budhwar Peth, Appa Balwant Chowk, Pune–411002, Ph.: 020-24475743
E-mail: pune.books@alliedpublishers.com
12 Prem Nagar, Ashok Marg, Opp. Indira Bhawan, Lucknow–226001, Ph.: 0522-2614253
E-mail: appltdlko@sify.com
Prarthna Flats (2nd Floor), Navrangpura, Ahmedabad–380009, Ph.: 079-26465916
E-mail: ahmbd.books@alliedpublishers.com
3-2-844/6 & 7 Kachiguda Station Road, Hyderabad–500027, Ph.: 040-24619079
E-mail: hyd.books@alliedpublishers.com
5th Main Road, Gandhinagar, Bangalore–560009, Ph.: 080-22262081
E-mail: bngl.books@alliedpublishers.com
1/13-14 Asaf Ali Road, New Delhi–110002, Ph.: 011-23239001
E-mail: delhi.books@alliedpublishers.com
17 Chittaranjan Avenue, Kolkata–700072, Ph.: 033-22129618
E-mail: cal.books@alliedpublishers.com
81 Hill Road, Ramnagar, Nagpur–440010, Ph.: 0712-2521122
E-mail: ngp.books@alliedpublishers.com
751 Anna Salai, Chennai–600002, Ph.: 044-28523938
E-mail: chennai.books@alliedpublishers.com

Website: www.alliedpublishers.com

ISBN 978-81-8424-895-1

Typeset by: Fortune Graphics

Published by Sunil Sachdev and printed by Ravi Sachdev at Allied Publishers Pvt. Ltd., Printing Division, A-104 Mayapuri Phase II, New Delhi-110064

PREFACE

This book is suitable to be a text for B.A./B.Sc. (Pass and Honours) and M.A./M.Sc. students of all Indian Universities and second and third year undergraduate students of the Universities of North America and Europe. Although this is not an exact translation but is written in the light of a book written in Oriya language by late Dr. Minaketan Das and published by the Bureau of Textbooks preparation and production, Orissa, Bhubaneswar.

I am thankful to Mrs. Das for providing me with a copy of the book and requesting me to write a book in English in the light of that book. This was the desire of Dr. Das, but he could not survive to author/co-author the book, or to see the first edition in print from. I thank professor G. Das and Professor Sribtsa Nanda for their help and encouragement which have always been source of inspiration.

The prerequisite for the book are basic concepts from set theory, abstract algebra and calculus/real analysis and we have included these concepts in the appendices to make the book self contained.

During the preparation of the book I have been benefited and made use of works of several authors including M.K. Das, Chandra Sekharan, Hardy and Wright, Pollard, M.M. Nanda, N. Tripathy, Apostal, Niven and Zuckerman (more of M.K. Das). I gladly take this opportunity to acknowledge my indebtedness to them.

I also thank Late Kaviraj Lambodar Das, my grandfather Late Upendra Nath Mishra and my uncle Sri Banshidhar Mishra who have helped and encouraged me in critical moment of my education. I am thankful to my wife who has suffered in isolation and looked after the family to enable me to complete the work.

Sudarsan Nanda

Dedicated to

Professor Minaketan Das

who served for the cause of Mathematics for over two decades but did not survive to see this book in print.

CONTENTS

CHAPTER 1

Prime Numbers

1.1. The Oldest and the most basic mathematical system is the Set N of all positive integers or natural numbers. Kronecker once said "God made the positive integers; all else is due to man." However, it was shown by the famous Italian mathematician G. Peano (1858–1932) that the positive integers can even more simply be defined by a set of axioms, known as Peano axioms. To describe these axioms, we first need to define the so-called Peano's successor function. This function, denoted by s, assigns to every positive integer the next larger one. Thus, $s : \mathrm{N} \to \mathrm{N}$ is defined by

$$s(n) = n + 1$$

In terms of the function s, a number of basic properties can be formulated which characterise the set N. These properties can be summarised as follows:

(*a*) $\mathrm{I} \in \mathrm{N}$;
(*b*) $n \in \mathrm{N} \Rightarrow s(n) \in \mathrm{N}$;
(*c*) there is no $n \in \mathrm{N}$ such that $s(n) = 1$;
(*d*) $s(n) = s(m) \Rightarrow n = m$;
(*e*) if $\mathrm{A} \subset \mathrm{N}$ such that $1 \in \mathrm{N}$, $s(n) \in \mathrm{N}$

whenever $n \in \mathrm{N}$, then $\mathrm{A} = \mathrm{N}$.

Axiom (*e*) leads to one of the most important principles of mathematics, known as the principle of mathematical induction.

We write N for the set $\{1, 2, 3, ...\}$ of all positive integers and Z for the set $\{..., -3, -2, -1, 0, 1, 2, ...\}$ of all integers. We write $a, b, c, ..., l, m, n, ..., x, y, z$, etc, for integers.

Given $a, b \in \mathrm{Z}$ $(b \neq 0)$ if there is $c \in \mathrm{Z}$ such that $a = bc$, then we say that b divides a and we write $b \mid a$. It is clear that

(*i*) if $c \mid b$ and $b \mid a$, then $c \mid a$,
(*ii*) if $c \mid b$ and $c \mid a$, then for any $m, n \in \mathrm{Z}$, we have $c \mid ma + nb$.

1.2. Positive integers can be divided into two classes: Prime numbers and Composite numbers.

Definition: $p \in N$ is said to be a Prime number if

(*i*) $p > 1$,

(*ii*) p is not divisible by any positive integer except 1 and p.

On the other hand, any positive integer greater than 1 which is not a prime is called a composite number.

Theorem 1. *Every positive integer greater than 1 is the product of (certain) prime numbers.*

Proof. Suppose that the given number is c. If it is a Prime number, then there is nothing to prove. If it is a composite number, then it is divisible by some integers greater than 1 and less than c. Let p_1 be the smallest of such integers. Then, p_1 is a prime number. For if p_1 is divisible by some positive integers $l > 1$ ($l \neq p_1$), then l will also divide c. But l being less than p_1, this contradicts the choice of p_1. Let $c = n_1 p_1$. Here, $1 < n_1 < c$. Now, n_1 is either a prime number or is divisible by a prime number p_2, *i.e.*, $n_1 = n_2 p_2$, where $1 < n_2 < n_1$. Similarly, n_2 is either prime or is divisible by a prime number p_3. Thus, $n_2 = n_3 p_3$, where $1 < n_3 < n_2$. Continuing in this way, we get a decreasing sequence $\{n_1, n_2, n_3, ...\}$ of positive integers each greater than 1 and less than c. Let n_k be the smallest of these. Now, n_k is a prime number ($= p_k$), for otherwise, if n_{k+1} divides n_k, then n_k will not be the smallest of $\{n_1, n_2, n_3, ...\}$. Hence, $n = p_1 p_2 p_3 ... p_k$ and this completes the proof.

Note that the prime numbers $p_1, p_2, ...$ need not be different from each other. If some of the p_m's are identical, then c will be represented as $c = p_1^{a_1} p_2^{a_2} ... p_m^{a_m}$ ($a_1 > 0$, $a_2 > 0$, ..., $p_1 < p_2 < ...$). We now prove

Theorem 2. (The Fundamental Theorem of Arithmetic). *The above representation of c is unique.*

Before proving this, we first prove Theorem 3.

Theorem 3. (Euclid's First Theorem). *If some prime number p divides ab, then p divides either a or b.*

Proof. Suppose that p does not divide a. We now show that p divides b.

Firstly; consider the set $\{pX + aY : X, Y \text{ are integers}\}$. Clearly, $pX + aY$ is an integer whenever X and Y are integers. Let q be the least positive integer which can be expressed in the above manner. So, $q = px_1 + ay_1$ for some $x_1, y_1, \in Z$. For some integers X and Y, if $pX + aY \in N$, then this is either greater than q or equal to q. Let

$$pX + aY = qs + r = (px_1 + ay_1)s + r$$

or,

$$p(X - x_1s) + a(Y - y_1s) = r$$

Now, r is of the form $px + ay$. If r would be positive, then it must be greater than or equal to q. But clearly, $r < q$. Therefore, $r = 0$. Thus, any integer which can be expressed in the from $pX + aY$ is divisible by q. Also $p = p.1 + a.0$. Thus p is divisible by q. But p is a prime number and hence $q = 1$ or $q = p$. Since q divides

$a = p.0 + a.1$, q cannot be equal to p. So $q = 1$.

Thus,

$$px_1 + ay_1 = 1$$

or

$$px_1b + ay_1b = b$$

Since p divides both px_1b and ay_1b, we conclude that p divides b.

Proof of Theorem 2

Let $p_1^{a_1} p_2^{a_2} \ldots p_k^{a_k} = c = q_1^{b_1} q_2^{b_2} \ldots q_j^{b_j}$.

Since q_1 divides $p_1^{a_1} p_2^{a_2} \ldots p_k^{a_k}$, q_1 must divide some p_t.

Again, since q_1 and p_t are prime numbers, $p_t = q_1$. Similarly, each p is equal to some q and each q is equal to some p. So $k = j$ and $p_1 = q_1, p_2 = q_2, \ldots, p_k = q_k$. If $a_1 < b_1$, then since

$$p_2^{a_2} p_3^{a_3} \ldots p_k^{a_k} = q_1^{b_1 - a_1} q_2^{b_2} \ldots q_k^{b_k},$$

q_1 will divide at least one of $p_2, p_3, p_4, \ldots, p_k$. So $a_1 < b_1$ and similarly $b_1 < a_1$ are impossible. Hence $a_1 = b_1$ and similarly, $a_2 = b_2$, $a_3 = b_3$, ... , $a_k = b_k$ and this completes the proof.

1.3. Let a and b be two given positive integers. Consider the set

$$\{ax + by : x \text{ and } y \text{ are integers}\}$$

Such a set (which is a ring with respect to usual addition and multiplication of integers) is called a *Number Ring*. Let d be the smallest positive integer which belongs to this set. Thus, there are integers x_1 and y_1 such that $d = ax_1 + by_1$ and if $ax + by$ is a positive integer, then this is either greater than or equal to d.

We now prove that d divides each element of the number ring and if some integer n divides a and b both, then n divides d.

Proof. Let $\quad a = dq + r$, where $0 \le r < d$.

So
$$\begin{aligned} r &= a - dq = a - (ax_1 + by_1)q \\ &= a(1 - qx_1) + b(-qy_1) \\ &= ax_1 + by_2 \end{aligned}$$

If r would be positive, then it would be greater than or equal to d. But this being impossible, $r = 0$. Thus, d divides a and similarly d divides b.

If there is some integer n which divides a and b both, then $a = nk$ and $b = nl$ for some integers k and l. So, $d = ax_1 + by_1 = (kx_1 + ly_1)n$ and hence $n \mid d$.

Note that d is said to be the *Highest Common Factor* (H.C.F.) of a and b and is written as (a, b).

It is clear that if $d = (a, b)$, in order that the equation $ax + by = n$ will have integer solutions d must divide n. On the other hand, if d divides n, then we can find atleast one pair of integers (x, y) for which $ax + by = n$.

If $(a, b) = 1$, then a and b are said to be *co-prime* and in this case, a and b do not have any common factor other than 1.

Similarly, for given m positive integers $a_1, a_2, \ldots, a_m$, we can find a positive integer d such that d divides each a_k and if any other positive integer n divides each a_k, then n must divide d. This d is called the H.C.F. of $a_1, a_2, \ldots, a_m$ and written as $d = (a_1, a_2, \ldots, a_m)$. As before we can find m integers $x_1, x_2, \ldots, x_m$ *for which*

$$d = a_1x_1 + a_2x_2 + \ldots + a_mx_m$$

We now prove this by induction on m. Firstly note that the result is true for $m = 2$. If $m > 2$, then observe that $(a_1, a_2, ..., a_m) = (a_1, a_2 ...,a_{m-1}), a_m)$. For if $d_{m-1} = (a_1, a_2 ..., a_{m-1})$ and $d_m = (a_1, a_2 ..., a_m)$, then $d_m \mid a_4$ $(i = 1, 2, ..., m)$. So $d_m \mid d_{m-1}$, but $d_m \mid a_m$. On the other hand, if $c \mid d_{m-1}$ and $c \mid a_m$, then $c \mid d_m$. So, $d_m = (d_{m-1}, a_m)$.

Now, assume that if $d_r = (a_1, a_2 ..., a_r)$, then there are integers $(y_1, ..., y_r$ for which $dr = a_1y_1 + ... + a_ry_r$ holds. Then, for two integers, Z_{r+1} and w_r, we have

$$
\begin{aligned}
d_{r+1} &= (d_r, a_{r+1}) = d_rw_r + a_{r+1}Z_{r+1} \\
&= a_1(y_1w_r) + a_2(y_2w_r) + ... + a_r(y_rw_r) + a_{r+1}Z_{r+1} \\
&= a_1z_1 + a_2z_2 + ... + a_rz_r + a_{r+1}Z_{r+1}
\end{aligned}
$$

This completes the proof.

1.4. The H.C.F. of two natural numbers a and b is obtained by *Euclid's algorithm* which is described below.

Firstly if $a = b$, then $(a, b) = (a, a) = a$. Assume that $a \neq b$, to be more specific, let $b < a$. Divide a by b and let r_1 be the remainder. Then, divide b by r_1 and let r_2 be the remainder. Now, divide r_1 by r_2 and let r_3 be the remainder. Continuing in this way, we get r_4, r_5,... such that $b > r_1 > r_2 > r_3 > ... > r_k > r_{k+1} > ...$ Since $\{r_1, r_2, ...\}$ is a decreasing sequence of integers, there is some integer n such that $r_{n+1} = 0$.

The above facts can be expressed by the following equations:

$$
\begin{aligned}
a - q_1b &= r_1, \\
a - q_2r_1 &= r_2, \\
r_1 - q_3r_2 &= r_3, \\
&\cdots\cdots\cdots \\
r_{n-2} - q_nr_{n-1} &= r_n, \\
r_{n-1} - q_{n+1}r_n &= 0
\end{aligned}
$$

Put $r_n = h$, it is clear from the last equation that $h \mid r_{n-1}$. Since $h \mid r_n$ and $h \mid r_{n-1}$, it follows that $h \mid r_{n-2}$. Similarly, it follows that $h \mid r_{n-3}, h \mid r_{n-4}, ..., h \mid r_2, h \mid r_1$. Thus $h \mid b$ and $h \mid a$.

On the other hand, if $d \mid a$ and $d \mid b$, then d divides $r_1 = a - q_1 b$. Again since $d \mid b$ and $d \mid r_1$, d divides $r_2 = b - q_2 r_1$. Similarly, it follows that d divides $r_2 = r_1 - q_3 r_2$. Continuing in this way, finally we observe that d divides $h = r_n = r_{n-2} - q_n\, r_{n-1}$.

Thus, $(a, b) = h$ and finally we observe that

$$\begin{aligned} h = r_n = r_{n-2} - q_n\, r_{n-1} &= r_{n-2} - q_n\,(r_{n-3} - r_{n-1}r_{n-2}) \\ &= (-q_n)r_{n-3} + (q_n q_{n-1} + 1)r_{n-2} \\ &= (-q_n)\, r_{n-3} + (q_n q_{n-1} + 1)\,(r_{n-4} - q_{n-2}\, r_{n-3}) \\ &= (q_n q_{n-1} + 1)\, r_{n-4} + (-\, q_n - q_{n-2} - q_n q_{n-1}\, q_{n-2})r_{n-3} \\ &= \ldots \qquad = a\,\mathrm{P}_n + b\,\mathrm{Q}_n \end{aligned}$$

For example, consider 741 and 1079. Then,

```
741)1079(1
     741
  --------
     338)741(2
         676
     -------
          65)338(5
             325
          ------
              13)65(5
                 65
              -----
                  0
```

Hence, $(741, 1079) = 13$. Further,

$$\begin{aligned} 13 &= 338 - 325 = 338 - 5(65) = 338 - 5\{741 - 2(338)\} \\ &= 11(338) - 5(741) = 11(1079 - 741) - 5(741) \\ &= 1079.\ (11) + 741.\ (-16) = a_1 x_1 + a_2 x_2, \end{aligned}$$

where, $x_1 = 11$ and $x_2 = -16$.

1.5. If two integers k and h are co-prime (*i.e.*, $(k, h) = 1$), then we can obtain two integers x and y such that $kx + hy = 1$. We now illustrate this by an example.

Let n be a given positive integer. Arrange all the fractions between 0 and 1 whose numerator and denominator are co-prime and whose denominator does not exceed n, in an increasing order. This is called the Farey Series and is denoted by F_n. Thus,

$$F_n = \left\{ \frac{p_1}{q_1}, \frac{p_2}{q_2}, \dots, \frac{p_s}{q_s} : (p_k, q_k) = 1, 0 < q_k \le n, \frac{p_r}{q_r} < \frac{p_{r+1}}{q_{r+1}} \right\}.$$

For example, if $(h, k) = 1$ and $0 \le h \le k \le n$, then $\frac{h}{k} \in F_n$.

Again, since $0 = \frac{0}{1}$ and $1 = \frac{1}{1}$, both 0 and 1 are in F_n.

For example,

$$E_5 = \left\{ \frac{0}{1}, \frac{1}{5}, \frac{1}{4}, \frac{1}{3}, \frac{2}{5}, \frac{1}{2}, \frac{3}{5}, \frac{2}{3}, \frac{3}{4}, \frac{4}{5}, \frac{1}{1} \right\}$$

We now prove that in a particular Farey Series,

(*i*) If $\frac{h'}{k'}$ is the element next to $\frac{h}{k}$, then $kh' - kh' = 1$ and

(*ii*) If $\frac{h''}{k''}$ is the element next to $\frac{h'}{k'}$, then $\frac{h'}{k'} = \frac{h+h''}{k+k''}$.

Proof of (*i*). Since h and k are co-prime, $kx - hy = 1$ has integer solutions. If x_0, y_0 are solutions of $kx - hy = 1$, then for any integer r, $x_0 + rh$, $y_0 + rk$ are also solutions. Choose r such that $n - k < y_0 + rk \le n$. For this r, let $x = x_0 + rh$ and $y = y_0 + rk$. For these x and y, $kx - hy = 1$ and hence, x and y are co-prime. Again, since $k > h$, we have $x < y$ (for if $x > y$, then $kx - hy > 1$).

Now, $(x, y) = 1$ and since $y \le n, \frac{x}{y}$ belongs to F_n. On the other hand,

$$\frac{x}{y} = \frac{h}{k} + \frac{1}{ky} > \frac{h}{k}$$

Therefore, $\frac{x}{y}$ is some element to the right of $\frac{h}{k}$ in F_n. We now prove

that $\frac{x}{y} = \frac{h'}{k'}$. Otherwise, if $\frac{x}{y} \ne \frac{h'}{k'}$ then $\frac{x}{y} > \frac{h'}{k'}$. But $k'x - h'y$ being an integer, we have $\frac{x}{y}\ \frac{h'}{k'}\frac{k'x-h'y}{k'y}\geq\frac{1}{k'y}$. Again, since $h'k - hk'$ is an integer, we have

$$\frac{h'}{k'}-\frac{h}{k} = \frac{h'k-hk'}{k'k}\geq\frac{1}{k'k}$$

So,
$$\frac{1}{ky}=\frac{kx-hy}{ky} = \frac{x}{y}-\frac{h}{k}=\left(\frac{x}{y}-\frac{h'}{k'}\right)+\left(\frac{h'}{k'}-\frac{h}{k}\right)$$

$$\geq \frac{1}{k'y}+\frac{1}{k'k}=\frac{k+y}{k'yk}>\frac{n}{k'ky}\geq\frac{1}{ky}$$

(Here, we have used $n - k < y_0 + rk = y \leq n$)

But $\frac{1}{ky}>\frac{1}{ky}$ is impossible and consequently, $\frac{x}{y}\neq\frac{h'}{k'}$ is impossible.

Thus $\frac{x}{y}=\frac{h'}{k'}$. But $(x, y) = 1 = (h', k')$. So, $x = h'$ and $y = k'$

Therefore, $kh' - k'h = kx - hy = 1$

Proof of (ii). By (*i*), we get

$$kh' - k'h = 1 \qquad \text{...(A)}$$

$$k'h'' - h'k'' = 1 \qquad \text{...(B)}$$

Multiplying h' with (A) and h with (B) and then adding, we get

$$h'(kh'' - kh'') = h + h'' \qquad \text{...(C)}$$

Again, multiplying k'' with (A) and k with (B) and adding, we get

$$k'(kh'' - hk'') = k + k'' \qquad \text{...(D)}$$

Now, dividing (C) and (D), we get

$$\frac{h'}{k'} = \frac{h+h''}{k+k''}$$

1.6. We now discuss the Lowest Common Multiple (L.C.M.) of two positive integers a and b.

Definition. If m is such an integer that (*i*) $a \mid m$, $b \mid m$ and (*ii*) if $a \mid c$, $b \mid c$, then $m \mid c$ and then m is said to be the L.C.M. of a and b and written as $m = [a, b]$. It is clear that there cannot be more than one integers satisfying the above two conditions. We now prove that $[a, b]. (a, b) = a.b$.

Proof. Let $(a, b) = d$, $a = a_1 d$, $b = b_1 d$, and $(a_1, b_1) = 1$. It now suffices to prove that $[a, b] = a_1 b_1 d = \dfrac{a.b}{d}$

Clearly, $a \mid ab_1 = a_1 b_1 d$ and $b \mid a_1 b = a_1 b_1 d$.

Secondly if $a \mid c$ and $b \mid c$, then $c = c_1$, $a = c_1 a_1 d$, $c = c'_1$ and $b = c'_1 b_1 d$,. So, $c_1 a_1 = c'_1 b_1$. Since $(a_1, b_1) = 1$, $b_1 \mid c_1$. Thus take $c_1 = kb_1$. So $c = c_1 a_1 d - kb_1 a_1 d$, *i.e.*, $a_1 b_1 d \mid c$, so $a_1 b_1 d = [a, b]$ and $ab = [a, b]. (a, b)$.

Definition. Let $a_1, a_2, \ldots a_k$ be given positive integers. Let m be such an integer that

(*i*) $a_1 \mid m$ for $i = 1, 2, \ldots, k$,

(*ii*) $a_1 \mid c$ for $i = 1, 2, \ldots, k$, then $m \mid c$.

Then, m is said to be the L.C.M. of $a_1, a_2, \ldots, a_k$ and is written as $m = [a_1, a_2, \ldots, a_k]$.

We will prove that if for $i = 1, 2, \ldots, k$.

$a_i = p_1^{x_{i1}}, p_2^{x_{i2}} \ldots p_r^{x_{ir}}$ (p_m being prime and $x_{ij} \geq 0$) and for $j = 1, 2, \ldots, r$, v_j is the greatest of all x_{ij} $(i = 1, 2, \ldots, k)$, then

$$[a_1, a_2, \ldots, a_k] = p_1^{v_1}\, p_2^{v_2} \ldots p_r^{v_r}.$$

Proof. Clearly for each i, $a_i \mid p_1^{v_1}, p_2^{v_2} \ldots p_r^{v_r}$. If for $i = 1, 2, \ldots, k$, $a_i \mid c$ and if for $i = s$, $x_{ij} = v_1$, then from $a_s \mid c$ we get $p_1^{v_1} \mid c$. Similarly, each one of $p_2^{v_2}, p_3^{v_3}, \ldots, p_r^{v_r}$ divides c and hence $p_1^{v_1}, p_2^{v_2} \ldots, p_r^{v_r} \mid c$.

1.7. In this section, we discuss some properties of prime numbers.

(*i*) **Euclid's Second Theorem.** *There are infinitely many prime numbers.*

Proof. Let p be a prime number. It is enough to show that there is a prime number greater than p.

Let, $$q=(2.3.5.7.11...p) + 1.$$

Now, q is not divisible by any one of 2, 3, 5, 7, 11,..., p. Therefore, every prime factor of q is greater than p, *i.e.*, either q in divisible by prime number greater than p or q itself is prime. This completes the proof.

It may be noted here that every prime number greater than 2 is an odd integer, so it can be expressed by $4n + 1$ or $4n + 3$.

(*ii*) There are infinitely many prime numbers which can be expressed in the form $4n + 3$.

Proof. Firstly, $3 = 4.0 + 3$ is a prime number.

If $Q = (2^2.3.5.7.11...p) - 1$, then Q can be expressed in the form $(4n + 3)$. If none of 2, 3, ..., p divide Q, so every prime factor of Q is greater than p. On the other hand, Q cannot be expressed as a product of prime numbers of the form $(4m + 1)$, since $(4n_1 + 1)$, $(4n_2 + 1)$ ... $(4n_k + 1) = (4n + 1)$. Thus, Q will be divisible by a prime number greater than p and expressible in the form $(4n + 3)$. Therefore, given a prime p there always exists a prime number greater than p and expressible in the form $(4n + 3)$ and this proves the assertion.

Similarly, it can be proved that there are infinitely many prime numbers which can be expressed in the form $(6n + 5)$ and we leave it as an exercise for the reader.

(*iii*) **Dirichlet's Theorem.** *If* $(a, b) = 1$, *then there are infinitely many prime numbers which can be expressed in the form* $(an + b)$.

We will not be able to prove this theorem in this book.

1.8. In this section, we shall discuss some useful symbols.

(*i*) **Greatest Integer Function.** For any number x we define $[x]$ as follows: If x is less than $(a + 1)$ and greater than or equal to a (where a is an integer), then $[x] = a$, *i.e.*, $[x]$ is the greatest integer not greater than x. For example, $[1.25] = 1$, $[3.7] = 3$, $[5.01] = 5$, $[-3.7] = -4$, $[-1.25] = -2$ and so on.

(*ii*) For two integers a and b, if $(a - b)$ is a positive integer, then a is said to be greater than b and written as $a > b$. Here, b is less than a and written as $b < a$.

(*iii*) If r is a negative integer, then $(-r)$ is a positive integer and is called the absolute value or r, Each positive integer is absolute value of itself. Absolute value of any number x, written as $|x|$, is defined as follows:

$$|x| = \begin{cases} x \text{ if } x \text{ is positive} \\ 0 \text{ if } x = 0 \\ -x \text{ if } x \text{ is negative} \end{cases}$$

(*iv*) If b is greater than or equal to a, we write $a \leq b$. If y is less than or equal to x, we write $x \geq y$.

EXERCISE

1. If for $i \neq j$, $(a_i, a_j) = 1$, prove that

$[a_1, a_2, ..., a_k] = a_1, a_2, ..., a_k$.

2. Prove that

$[a_1, a_2, ..., a_k] = [[a_1, a_2, ..., a_{k-1}], a_k]$

3. If $(a, b) = 1$ and $c \mid a$, prove that $(c, b) = 1$.

4. Prove that $(ma, mb) = m(a, b)$.

5. If $(a, b) = 1$, then prove that $(ac, b) = (c, b)$.

6. In order that $(a, n_1, n_2, ..., n_k) = 1$, it is necessary and sufficient that $(a, n_1) = (a, n_2) = ... = (a, n_k) = 1$.

7. If u_m is the smallest of $x_{m_1}, x_{m_2}, ..., x_{mr}$, then show that

$(a_1, a_2, ..., a_k) = p_1^{u_1} p_2^{u_2} ... p_r^{ur}$.

CHAPTER 2

Congruence

2.1. Let m be a positive integer and $m \mid (x - a)$. Then x is said to be congruent to a mod m and written as $x \equiv a \pmod{m}$. On the other hand, if $m + (x - a)$, then x is said to be incongruent to a mod m and written as $x \not\equiv a \pmod{m}$.

Let $x \equiv a \pmod{m}$. Then, a is said to be a residue of x mod m. Again, if $0 \leq a \leq m - 1$, then a is said to be the *least residue* of x mod m.

For a given residue, the set of all integers congruent to this mod m is called the *class of residues* mod m. The difference of any two elements of this class is divisible by m.

Example. Let $m = 4$, $a = 3$. Then if $x \equiv 3 \pmod{4}$, $4 \mid (x - 3)$, *i.e.*, $x - 3 = 4q$ or. $x = 4q + 3$. By taking different integer values of q, we obtain the set of values of x:

$$\{..., -17, -13, -9, -5, -1, 3, 7, 11, 15, 19,...\}.$$

We denote this set by $\overline{3}$.

Similarly,

$$\overline{0} = \{..., -4, 0, 4, 8, ...\}.$$

$$\overline{1} = \{..., -7, -3, 1, 5, 9, ...\}$$

$$\overline{2} = \{..., -6, -2, 2, 6, 10, ...\}.$$

Similarly, if $m = 5$, we get

$$\overline{0} = \{..., -15, -10, -5, 0, 5, 10, ...\}$$

$$\overline{1} = \{..., -14, -9, -4, 1, 6, 11, ...\}$$

$$\overline{2} = \{..., -13, -8, -3, 2, 7, 12, ...\}$$

$$\overline{3} = \{..., -12, -7, -2, 3, 8, 13, ...\}$$

$$\overline{4} = \{..., -11, -6, -1, 4, 9, 14, ...\}$$

It is clear that for a given m, the m number of residue classes are as follows:

$$\overline{0} = \{..., -3m, -2m, -m, 0, m, 2m, 3m, ...\}$$

$$\overline{1} = \{..., -3m + 1, -2m + 1, -m + 1, 1, m + 1, 2m + 1, 3m + 1, ...\}$$

$$\overline{2} = \{..., -3m + 2, -2m + 2, -m + 2, 2, m + 2, 2m + 2, 3m + 2, ...\}$$

....

$$\overline{r} = \{..., -3m + r, -2m + r, -m + r, r, m + r, 2m + r, 3m + r, ...\}$$

....

$$\overline{m-1} = \{..., -2m - 1, -m - 1, -1, m - 1, 2m - 1, 3m - 1, ...\}.$$

The set consisting of $\overline{0}, \overline{1}, \overline{2}, ..., \overline{m-1}$ or the set formed by choosing one representative from each of the above sets is called a complete system of incongruent residues.

Some properties of congruences: It is clear that for a given m,

(*i*) $a \equiv a$ (reflexivity),

(*ii*) $a \equiv b \Rightarrow b \equiv a$ (symmetry),

(*iii*) $a \equiv b$, $b \equiv c \Rightarrow a \equiv c$ (transitivity),

(*iv*) $a \equiv b$ and k is any integer $\Rightarrow ka \equiv kb$,

(*v*) $a \equiv b$ and $a' \equiv b' \Rightarrow a + a' = b + b'$ and $aa' \equiv bb'$.

Further, for any positive integer n, $a^n \equiv b^n$.

Note that for a given m, the relation. "Congruence mod m" is an equivalence relation and hence defines a partition in the set of all integers.

2.2. Linear Congruence

If $ka \equiv ka'$ (mod m), then it is not necessarily true that $a = a'$ (mod m). For example, $2.8 \equiv 2.5$ (mod 6), but $8 \neq 5$ (mod 6). But this property holds under certain conditions.

Theorem 4. *If k and m are coprime and if $ka \equiv kb$ (mod m)*, then $a \equiv b$ *(mod m)*.

Proof. Let $a \neq b$ (mod m). Then for some r ($0 < r < m$), $a - b = mq + r$. Again since $(k, m) = 1$, $kx + my = 1$ has an integer solution $x = x_0$ and $y = y_0$.

So,

$$krx_0 + mry_0 = r \quad \text{or, } r = m(y_0r) + x_0(ka - kb - kmq).$$

But it is given that $ka = kb$ (mod m). So, $r = 0$ (mod m), *i.e.*, $m \mid r$. But this is impossible since $0 < r < m$ and this completes the proof of the theorem.

Theorem 5. *If $(k, m) = d$ and if $ka \equiv kb$ (mod m), then $a \equiv b$* $\left(\text{mod } \frac{m}{d}\right)$.

Proof. Let, $k = k_1d$, $m = m_1d$ and $(k_1, m_1) = 1$.

Since $\dfrac{ka - kb}{m} = \dfrac{k_1(a-b)}{m_1}$, we have

$$k_1a = k_1b \text{ (mod } m_1).$$

Since $(k_1, m_1) = 1$ from theorem 4, it follows that $a \equiv b$ (mod m_1).

Theorem 6. *If $(k, m) = 1$, then $kx \equiv l$ (mod m) has exactly one solution.*

Proof. In order that $(kx - l)$ is divisible by m, $kx - l = my$ or, $kx - my = l$. The smallest positive integer belonging to the set $\{kx - my : x \text{ and } y \text{ are integers}\}$ is 1. If $kx_0 - my_0 = 1$, then for $x = lx_0$, $kx - my = l$, *i.e.*, $kx = l$ (mod m). On the other hand, let x_1 and x_2 be two solutions of the given congruence. Then since $kx_1 \equiv l \equiv kx_2$ (mod m) by Theorem 4 it follows that $x_2 \equiv x_2$ (mod m). Thus if for mod m, $kx_1 \equiv l$, $kx_2 \equiv l$, $kx_3 \equiv l$, ... $kx_r \equiv l$, ..., then $x_1 \equiv x_2 \equiv x_3 \equiv ... \equiv x_r \equiv ...$ (mod m). Hence, all the solutions of $kx \equiv l$ (mod m) form only one Residue Class. Since congruence mod m is an equivalence relation, each Residue class will be considered as a single element and therefore, $kx \equiv l$ (mod m) has only one solution.

For mod m, the set $\{\overline{0}, \overline{1}, \overline{2}, ..., \overline{(m-1)}\}$ will be denoted by Z_m. Theorem 6 essentially says that the set of all solutions of $kx \equiv l$ (mod m) is a single element of Z_m.

Theorem 7. *Let* $(k, m) = d$. *In order that* $kx \equiv l$ *(mod m) has a solution it is necessary that* $d \mid l$. *Conversely, if* $d \mid l$, *then the above congruence has d and only d different solutions.*

Proof. Suppose that $kx \equiv l$ (mod m) has a solution, *i.e.*, for some integers x_0 and y_0, $kx_0 - l = my_0$. Since d divides k and m, d will divide $kx_0 - my_0 = l$.

Conversely, let $d \mid l$. Put $k = k'd$, $m = m'd$ and $l = l'd$. Then $kx - l = my$ can be written as $k'x - l' = m'y$. Since $(k', m') = 1$, from theorem 6, it follows that $k'x \equiv l'$ (mod m') has exactly one solution. If this solution is given by $x \equiv t$ (mod m'), then the solutions of $k'x \equiv l'$ (mod m') are given by

$$\{..., -2m' + t, -m' + t, t, m' + t, 2m' + t, ...\}.$$

So, if $kx \equiv l$ (mod m), then $k'x \equiv l'$ (mod m') or, $x \equiv t$ (mod m') or, $x = t + m'y$. Now, if $t + m'y = t + m'z$ (mod m), then m will divide $m'(y - z)$, *i.e.*, $m'(y - z) = mw = m'dw$ or, $(y - z) = d.w$. Thus d will divide $(y - z)$. We have now proved that if $t + m'y = t + m'z$ (mod m), then $y \equiv z$ (mod d) or $z = y + nd$. Therefore, for mod m,

$$x \equiv t + nd,\ t + m' + nd,\ t + 2m' + nd...,$$
$$t + (d - 1)m' + nd\ (n = 0, \pm 1, \pm 2, ...).$$

The residue class mod m is given by

$$\{\bar{t}, \overline{t + m^1}, \overline{t + 2m^1}, ..., \overline{t + (d-1)\, m^1}\}.$$

Example 1. Suppose $m = 7$ and we want to solve $6x \equiv 2$ (mod 7). The possible values of x are congruent to 0, 1, 2, 3, 4, 5 or 6 (mod 7) and

$$6.\ \{0, 1, 2, 3, 4, 5, 6\} = \{0, 6, 5, 4, 3, 2, 1\}$$

respectively. Therefore, the solution of $6x \equiv 2$ (mod 7) is given by $x \equiv 5$ (mod 7).

Example 2. Let $m = 10$. Since 3 is not divisible by $(6, 10) = 2$, the congruence $6x \equiv 3 \pmod{10}$ has no solutions.

Example 3. Similarly, it can be seen that $6x \equiv 8 \pmod{15}$ has no solutions.

Example 4. Since 9 is divisible by $(6, 15) = 3$, the congruence $6x \equiv 9 \pmod{15}$ has solutions. The possible values of x are $\overline{0}, \overline{1}, \overline{2}, \overline{3}, \overline{4}, \overline{5}, \overline{6}, \overline{7}, \overline{8}, \overline{9}, \overline{10}, \overline{11}, \overline{12}, \overline{13}, \overline{14}$. By multiplying 6 with each of the values we respectively obtain $\overline{0}, \overline{6}, \overline{12}, \overline{3}, \overline{9}, \overline{0}, \overline{6}, \overline{12}, \overline{3}, \overline{9}, \overline{0}, \overline{6}, \overline{12}, \overline{3}, \overline{9}$. Hence the solution of $6x \equiv 9 \pmod{15}$ are $\overline{x} = \overline{4}, \overline{9}, \overline{14}$.

or $\qquad x = 4, 9, 14 \pmod{15}$

Example 5. Consider

$$11799x \equiv 8715 \pmod{1081}.$$

By taking $m = 1081$, we see that $11799 \equiv 989 \pmod{m}$ and $8715 \equiv 67 \pmod{m}$. So, if the given congruence has a solution, we must have $989x \equiv 67 \pmod{1081}$. But for $k = 989$, $m = 1081$, we have $d = (k, m) = (989, 1081) = 23$. This d does not divide 67 and hence, it follows from Theorem 7 that the given congruence has no solutions.

Example 6. We now determine the common solution of $x \equiv 5 \pmod{8}$ and $x \equiv 4 \pmod{9}$. From the first one, we get $x = 5 + 8t$. Putting this value in the second one, we get $5 + 8t \equiv 4 \pmod{9}$ or, $8t \equiv -1 \pmod{9}$. By multiplying the last one with 8 and using the fact that $8 \times 8 \equiv 1 \pmod{9}$, we get $64t \equiv -8 \pmod{9}$ or, $t \equiv -8 \equiv 1 \pmod{9}$. So, $t = 1 + 9r$ and $x = 5 + 8t = 5 + 8(1 + 9r) = 13 + 72r$.

Example 7. Proceeding as earlier one can show that $x \equiv 5 \pmod{8}$ and $x \equiv 4 \pmod{18}$ have no common solutions.

By the above procedure it is possible, but unpleasant, to think of the common solutions of more than two congruences. Therefore, we feel the necessity of a formula known as Chinese Remainder Theorem.

Chinese Remainder Theorem: *Let m_1, m_2, m_3, ..., m_n be integers each greater than 1 and coprime to each other. Then, the following simultaneous congruences*

$$x = a_1 \text{ (mod } m_1)$$
$$x = a_2 \text{ (mod } m_2)$$
$$\ldots\ldots\ldots\ldots\ldots\ldots$$
$$x = a_n \text{ (mod } m_n)$$

*has **a** solution x_0 and for any general solution X,*

$$X \equiv x_0 \text{ (mod } m_1\ m_2 \ldots m_n).$$

Proof. Let $m = m_1 m_2 \ldots m_n$ and for $k = 1, 2, \ldots, n$, $m'k = \frac{m}{m_k}$ $= m_1 m_2 \ldots m_{k-1}\, m_{k+1} \ldots m_n$. Given that $m_1, m_2, \ldots m_n$ are coprimes to each other. Hence, for each j, $(m_j', m_j) = 1$ and we can choose integers x_j and y_j such that $m_j' x_j + m_j y_j = 1$. Since $m_j' x_j = m_1 m_2 \ldots m_{j-1} m_{j+1} \ldots m_n x_j$, it follows that

$$m_j' x_j \equiv 1 \text{ (mod } m_j)$$

and $$m_j' x_j \equiv 0 \text{ (mod } m_i i \neq j).$$

So $$m_j' x_j a_j \equiv a_j \text{ (mod } m_j)$$

and $$m_j' x_j a_j \equiv 0 \text{ (mod } m_i i \neq j),$$

That is, if, $x_0 = m_1' x_1 a_1 + m_2' x_2 a_2 + \ldots + m_j' x_j a_j + \ldots + m_n' x_n a_n$, then for $k = 1, 2, \ldots, n$, $x_0 \equiv a_k$ (mod m_k). If y_0 is some other solution of the given simultaneous congruences, then $y_0 \equiv a_k$ (mod m_k) ($k = 1, 2, \ldots, n$) and hence $y_0 \equiv x_0$ (mod m_k). Thus each m_k ($k = 1, 2, \ldots, n$) divides $(y_0 - x_0)$. But $m_1, m_2, \ldots, m_n$ are coprime to each other. Therefore, $m_1 m_2 m_3 \ldots m_n$ will divide $(y_0 - x_0)$, *i.e.*, $y_0 - x_0 = mq$. Hence, the general (common) solution of the given simultaneous congruences is given by

$$x \equiv x_1 a_1 m'_1 + x_2 a_2 m'_2 + \ldots + x_n a_n m_n' + km$$

Example 1. Find the solution of

$$x \equiv 1 \text{ (mod 2)},\ x \equiv 2 \text{ (mod 3)},\ x \equiv 3 \text{ (mod 5)},\ x \equiv 5 \text{ (mod 7)}.$$

Here, $m_1 = 2$, $m_2 = 3$, $m_3 = 5$, $m_4 = 7$ are co-prime and $m = m_1 m_2 m_3 m_4 = 2.\ 3.\ 5.\ 7 = 270$. Again,

$$m_1' = \frac{m}{m_1} = 105,\ m_2' = \frac{m}{m_2} = 70,\ \ m_3' = \frac{m}{m_3} = 42,\ m_4' = \frac{m}{m_4} = 30.$$

Again, in order that $105x_1 \equiv 1 \pmod 2$, $70x_2 \equiv 1 \pmod 3$, $42x_3 \equiv 1 \pmod 5$ and $30x_4 \equiv 1 \pmod 7$, it is sufficient that $1.x_1 \equiv 1 \pmod 2$, $1.x_2 \equiv 1 \pmod 3$, $2\,x_3 \equiv 1 \pmod 5$ and $2.x_4 \equiv 1 \pmod 7$. So we can take $x_1 \equiv 1$, $x_2 \equiv 2$, $x_3 \equiv 3$ and $x_4 \equiv 4$. Now, since $a_1 \equiv 1$, $a_2 \equiv 2$, $a_3 \equiv 3$ and $a_4 \equiv 5$, the common solution of the given simultaneous congruences is

$$
\begin{aligned}
x &= x_1a_1m_1' + x_2a_2m_2' + x_3a_3m_3' + x_4a_4m_4' + km \\
&= (1)(1)\,(105) + (1)(2)\,(70) + (3)(3)\,(42) + (4)(5)\,(30) + k(210) \\
&= 1223 + 210k
\end{aligned}
$$

and among these, the smallest positive solution is $x_0 = 173$.

Example 2. In the following simultaneous congruences $x \equiv 6 \pmod{17}$, $x \equiv 17 \pmod{24}$, $x \equiv 13 \pmod{33}$, 24 and 33 are not co-primes. But if $x \equiv 17 \pmod{24}$, $(x - 17)$ is divisible by 24. So $(x - 17)$ is divisible by 3 and 8. Similarly, if $x \equiv 13 \pmod{33}$, $(x - 13)$ is divisible by 3 and 11. Hence, the given congruences can be written as:

$x \equiv 6 \pmod{17}$, $x \equiv 17 \pmod 3$, $x \equiv 17 \pmod 8$, $x \equiv 13 \pmod 3$ and $x \equiv 13 \pmod{11}$. If second and fourth hold, then $17 \equiv 13 \pmod 3$, but this is impossible. So, the given simultaneous congruences have no solution.

Example 3. The following simultaneous congruences $x \equiv 5 \pmod{18}$, $x \equiv -1 \pmod{24}$, $x \equiv 17 \pmod{33}$ can be expressed as:

$x \equiv 1 \pmod 2$, $x \equiv 5 \pmod 9$, $x \equiv -1 \pmod 3$, $x \equiv -1 \pmod 8$, $x \equiv 2 \pmod 3$, $x \equiv 6 \pmod{11}$

In this case, if $x \equiv 5 \pmod 9$, then $x \equiv -1 \pmod 3$ and $x \equiv 2 \pmod 3$. So, we discard the last two. Similarly if $x \equiv -1 \pmod 8$, then $x \equiv 1 \pmod 2$ and we discard the last one. That is, we consider $x \equiv 5 \pmod 9$, $x \equiv -1 \pmod 8$ and $x \equiv 6 \pmod{11}$, By using Chinese Remainder Theorem, we get the solution as

$$x = 743 + 792k \text{ or, } x \equiv 743 \pmod{792}.$$

Example 4. Consider

$$2x \equiv 4 \pmod{6} \text{ and } 5x \equiv 3 \pmod{11}.$$

The solution of the first one is $x \equiv 2$ or $5 \pmod{6}$ and that of the second one is $x \equiv 5 \pmod{11}$. So, the solution of the given congruences is the solution of

(*a*) $x \equiv 2 \pmod{6}$, $x \equiv 5 \pmod{11}$

or of

(*b*) $x \equiv 5 \pmod{6}$, $x \equiv 5 \pmod{11}$.

By Chinese Remainder theorem, we observe that the unique solution of (*a*) is $x \equiv 38 \pmod{66}$ and that of (*b*) is $x \equiv 5 \pmod{66}$. So, the two distinct solutions of the given simultaneous congruences are

$$x \equiv 5 \pmod{66} \text{ and } x \equiv 38 \pmod{66}.$$

Example 5. Consider $4x \equiv 6 \pmod{10}$ and $9x \equiv 15 \pmod{21}$. From the first one we get $x \equiv 4$ or $9 \pmod{10}$ and from the second one $x \equiv 4$ or 11 or $18 \pmod{21}$. Therefore, the solutions will be obtained from the solutions of the following six congruences:

(*i*) $x \equiv 4 \pmod{10}$, $x \equiv 4 \pmod{21}$
(*ii*) $x \equiv 9 \pmod{10}$, $x \equiv 4 \pmod{21}$
(*iii*) $x \equiv 4 \pmod{10}$, $x \equiv 11 \pmod{21}$
(*iv*) $x \equiv 9 \pmod{10}$, $x \equiv 11 \pmod{21}$
(*v*) $x \equiv 4 \pmod{10}$, $x \equiv 18 \pmod{21}$
(*vi*) $x \equiv 9 \pmod{10}$, $x \equiv 18 \pmod{21}$

By Chinese Remainder theorem, their solutions are respectively:

(*i*) $x \equiv 4 \pmod{210}$ (*ii*) $x \equiv 109 \pmod{210}$
(*iii*) $x \equiv 74 \pmod{210}$ (*iv*) $x \equiv 179 \pmod{210}$
(*v*) $x \equiv 144 \pmod{210}$ (*vi*) $x \equiv 39 \pmod{210}$

That is, the solution of the given two congruences is

$$x \equiv 4, 39, 74, 109, 144, 179 \pmod{210}$$

EXERCISE

1. Solve the following congruences:

(*i*) $158x \equiv -22 \pmod{194}$ (*ii*) $194x \equiv -22 \pmod{158}$

(*iii*) $84x \equiv 156 \pmod{605}$ (*iv*) $84x \equiv 156 \pmod{438}$

(*v*) $9x \equiv 6 \pmod{24}$ (*vi*) $12x \equiv 14 \pmod{26}$

(*vii*) $1001x \equiv 433 \pmod{4845}$ (*viii*) $115x \equiv 87 \pmod{253}$.

2. Solve the following simultaneous congruences:

(*i*) $x \equiv 3 \pmod 7$, $x \equiv -2 \pmod{11}$

(*ii*) $x \equiv -2 \pmod{11}$, $x \equiv 8 \pmod{22}$

(*iii*) $x \equiv 3 \pmod 7$, $x \equiv 7 \pmod{12}$, $x \equiv 8 \pmod{13}$

(*iv*) $x \equiv 2 \pmod 7$, $x \equiv -2 \pmod{11}$, $x \equiv 8 \pmod{13}$, $x \equiv -3 \pmod{17}$

(*v*) $x \equiv -2 \pmod{11}$, $x \equiv 8 \pmod{13}$, $x \equiv 9 \pmod{14}$, $x \equiv 1 \pmod{15}$, $x \equiv -3 \pmod{17}$

(*vi*) $x \equiv 3 \pmod 7$, $x \equiv -2 \pmod{11}$, $x \equiv 8 \pmod{13}$, $x \equiv -3 \pmod{17}$, $x \equiv 5 \pmod{18}$

(*vii*) $x \equiv 3 \pmod 7$, $x \equiv 9 \pmod{14}$

(*viii*) $x \equiv 7 \pmod{12}$, $x \equiv 9 \pmod{14}$

(*ix*) $x \equiv 7 \pmod{12}$, $x \equiv 1 \pmod{15}$

(*x*) $x \equiv 7 \pmod{12}$, $x \equiv 5 \pmod{18}$

(*xi*) $x \equiv 9 \pmod{14}$, $x \equiv 5 \pmod{18}$

(*xii*) $x \equiv 1 \pmod{15}$, $x \equiv 5 \pmod{18}$

(*xiii*) $x \equiv 7 \pmod{12}$, $x \equiv 9 \pmod{14}$, $x \equiv 1 \pmod{15}$

(*xiv*) $x \equiv 9 \pmod{14}$, $x \equiv -3 \pmod{17}$, $x \equiv 5 \pmod{18}$.

3. If $(a, m) = d$ divides b and if $a = da'$, $b = db'$ $m = dm'$, then show that although $a'x = b' \pmod{m'}$, $ax = b \pmod m$ are one and the same, the first one has d solutions whereas the second one has only one solution. How is it possible?

4. Solve the following simultaneous congruences:

$3x \equiv 5 \pmod{22}$, $11x \equiv 3 \pmod{28}$, $5x \equiv 89 \pmod{99}$.

5. Solve the following simultaneous congruences:

   $9x \equiv 6 \pmod{24}$, $12x \equiv 14 \pmod{26}$.

6. Find the smallest number which when divided by 3, 5 and 7, the remainders are 2, 3 and 2, respectively.

7. Show that $x \equiv a \pmod{m}$ and $x \equiv b \pmod{n}$ have one and the same solution if and only if (m, n) divides $(a - b)$.

8. If $(m_1, m_2) = 1$ and $m = m_1m_2$, then prove that the set of all solutions of $ax \equiv b \pmod{m}$ is same as the set of all solutions of $ax \equiv b \pmod{m_1}$ and $ax \equiv b \pmod{m_2}$.

Answers

1. (*i*) $x \equiv 6, 103 \pmod{194}$
   (*ii*) $x \equiv 74, 153 \pmod{158}$
   (*iii*) $x \equiv 434 \pmod{605}$
   (*iv*) $x \equiv 54, 172, 200, 346, 419 \pmod{438}$
   (*v*) $x \equiv 6, 14, 22 \pmod{24}$
   (*vi*) $x \equiv 12, 25 \pmod{26}$
   (*vii*) no solution.
2. (*i*) $x \equiv 31 \pmod{77}$, (*iii*) $x \equiv 619 \pmod{1092}$, (*vii*) no solution
   (*ix*) $x \equiv 31 \pmod{60}$, (*x*) no solution, (*xiii*) $x \equiv 331 \pmod{420}$.
3. $x \equiv 97 \pmod{2^2.\ 3^2.\ 7.\ 11}$.
4. There are three solutions mod (24, 26).
5. 23.

Application of Chinese Remainder Theorem: Let $a_0, a_1, \ldots, a_n$ *be integers and let m be a positive integer. We now try to solve*

$$a_0 + a_1x + a_2x^2 + \ldots + a_nx^n \equiv 0 \pmod{m}$$

Firstly, we illustrate this through some examples.

Example 1. If $x^3 - x + 4 \equiv 0 \pmod{7^2}$, then what is the value of x?

From the given congruence, we obtain

$$x^3 - x + 4 \equiv 0 \pmod{7}$$

or, $$(x - 3)(x^2 + 3x + 1) \equiv 0 \pmod{7}$$

Take $x - 3 = 7t$ and write the given congruence as

$$(3 + 7t)^3 - (3 + 7t) + 4 \equiv 0 \pmod{7^2}$$

From this, it follows that

$$28 + 182t + 431t^2 + 343t^3 \equiv 0 \pmod{7^2}$$

or, $$28 + 35t \equiv 0 \pmod{7^2}$$

So $28 + 35t = 7(4 + 5t)$ is divisible by 7^2, *i.e.*, $(4 + 5t)$ is divisible by 7, or $4 + 5t \equiv 0 \pmod 7$

or, $5t \equiv -4 \equiv 3 \pmod 7$ or $15t \equiv 9 \pmod 7$ or $t \equiv 2 \pmod 7$.

So, $t = 2 + 7u$ or $x = 3 + 7t = 3 + 7(2 + 7u) = 17 + 7^2u$

or, $x \equiv 17 \pmod{7^2}$.

Example 2. $x^2 - x + 2 \equiv 0 \pmod{7^2}$

Observe that $x^2 - x + 2 \equiv 0 \pmod 7$. So, $x^2 - x + 2 \equiv x^2 - 8x + 2 \equiv x^2 - 8x + 16 = (x - 4)^2 \equiv 0 \pmod 7$ or $x - 4 = 7t$. Putting this value of x in the given congruence, we get

$$(4 + 7t)^2 - (4 + 7t) + 2 \equiv 0 \pmod{7^2}$$

or, $$49t^2 + 49t + 14 \equiv 0 \pmod{7^2}$$

or, $$14 \equiv 0 \pmod{7^2}$$

But this is impossible and hence the given congruence has no solution.

Example 3. $x^2 + x + 1 \equiv 0 \pmod{7^2}$.

Firstly, in order that $x^2 + x + 1 \equiv 0 \pmod 7$, we must have

$$x^2 + x + 1 \equiv x^2 - 6x + 8 = (x - 2)(x - 4) \pmod 7$$

or, $$(x - 2)(x - 4) \equiv 0 \pmod 7$$

So, $$x \equiv 2 \text{ or } 4 \pmod 7.$$

(A) By taking $x \equiv 2 \pmod 7$ or, $x = 2 + 7t$ in the original equation, we have

$$(2 + 7t)^2 + (2 + 7t) + 1 \equiv 0 \pmod{7^2}$$

or, $$7(1 + 5t) \equiv 0 \pmod{7^2}$$

or, $$1 + 5t \equiv 0 \pmod 7$$

or, $3 + 15t \equiv 0 \pmod 7$ or, $t \equiv 4 \pmod 7$

or, $t = 4 + 7u$ or,

$$x = 2 + 7(4 + 7u)$$

$$= 30 + 7^2u$$

or, $x \equiv 30 \pmod{7^2}$

(B) By taking $x \equiv 4 \pmod 7$ or, $x = 4 + 7s$ in the original equation, we get

or, $(4 + 7s)^2 + (4 + 7s) + 1 \equiv 0 \pmod{7^2}$

or, $7(3 + 9s) \equiv 0 \pmod{7^2}$

or, $3 + 9s \equiv 0 \pmod 7$

or, $12 + s \equiv 0 \pmod 7$ or, $s \equiv 2 \pmod 7$

So $s = 2 + 7u$ or, $x = 4 + 7(2 + 7u) = 18 + 7^2u$

So, $x \equiv 18 \pmod{7^2}$.

Hence, the solution of the given congruence is

$$x \equiv 18, 30 \pmod{7^2}.$$

Example 4. $x^2 + x + 1 \equiv 0 \pmod{7^3}$.

Since $x^2 + x + 1 \equiv 0 \pmod{7^2}$, according to the above discussions,

$$x \equiv 18, 30 \pmod{7^2}.$$

(A) If $x \equiv 18 \pmod{7^2}$, then $x = 18 + 7^2u$ and from the given congruence we have

$$(18 + 7^2u)^2 + (18 + 7^2u + 1) \equiv 0 \pmod{7^3}$$

or, $7^2(7 + 37u + 7^2u^2) \equiv 0 \pmod{7^3}$

or, $7 + 37u + 7^2u^2) \equiv 0 \pmod 7$

or, $37u \equiv 0 \pmod 7$ or, $u \equiv 0 \pmod 7$

and $x = 18 + 7^2u = 18 + 7^2(0 + 7h) = 18 + 7^3h$

or, $x \equiv 18 \pmod{7^3}$

(B) If $x \equiv 30 \pmod{7^2}$, then $x = 30 + 7^2u$ and from the given congruence, we have

$$(30 + 7^2v)^2 + (30 + 7^2v) + 1 \equiv 0 \pmod{7^3}$$

or, $$7^2 (19 + 61v) \equiv 0 \pmod{7^3}$$

or, $$19 + 61v \equiv 0 \pmod 7$$

or, $$5 + 5v \equiv 0 \pmod 7$$

or, $$1 + v \equiv 0 \pmod 7$$

or, $$v \equiv 6 \pmod 7$$

So $$x \equiv 30 + 7^2v = 30 + 7^2(6 + 7k)$$

i.e., $$x \equiv 324 \pmod{7^3}.$$

From the above examples, the method of solution of

$$f(x) = a_0x^n + a_1x^{n-1} + a_2x^{n-2} + \dots + a_n \equiv 0 \pmod{p^r},$$

for any prime p, is evident.

Let $s < r$ and let $x \equiv c_s \bmod (p^s)$ be a solution of $f(x) \equiv 0, \pmod{p^s}$. By taking $C_{s+1} = C_s + tp^s$, we get

$$f(C_{s+1}) = f(C_s + tp^s) = f(C_s) + tp^s f(C_s) + (tp^s)^2 \text{ I } (C_s\ tp^3).$$

Here, $f(C_s) = a_0 n\ C_s^{\,n-1} + a_1 (n-1)\ C_s^{n-2} + \dots a_{n-1}$ is an integer and also $I(C_s, tp^s)$ is an integer. Again if $s \geq 1$, then $2s \geq s + 1$. Now since $f(c_s) \equiv 0 \pmod{p^s}$ by taking $f(c_s) = k.p^s$, we get

$$f(c_{s+1}) = p^s (p + t f'(c_s) + t^2p^{2s} \text{ I } (c_s, tp^r))$$

or, $$f(c_{s+1}) = p^s (k + t f'(c_s)) \pmod{p^{s+1}}$$

and we obtain the value of t from $k + tf'(c_s) \equiv 0 \pmod p$

Then $f(c_{s+1}) \equiv 0 \pmod{p^{s+1}}$, *i.e.,*

if $x \equiv c_s \pmod{p^s}$ is a solution of $f(x) \equiv 0 \pmod{p^s}$, then

$$c_{s+1} = c_s + tp^s \text{ is a solution of } f(x) \equiv 0 \pmod{p^{s+1}}.$$

The value of t is obtained, in this case, from

$$tf'(c_s) \equiv -k = -\frac{f(c_s)}{p^s} \pmod p$$

or, $tf'(c_s) = -\dfrac{f(c_s)}{p^s} \pmod p$ and for this t, $f(c_{s+1}) \equiv 0 \pmod{p^{s+1}}$.

If $f'(c_s) \equiv 0 \pmod{p}$ and $\dfrac{f(c_s)}{p^s} \equiv 0 \pmod{p}$, then the above congruence is true for any t and so t can be choosen from $t = 0, 1, 2, ..., p - 1$. Hence if $f'(c_s) \not\equiv 0 \pmod{p}$, then t has one value and if $f'(c_s) \equiv p^{-s} f(c_s) \equiv 0 \pmod{p}$, then t has p values. On the other hand, if $f'(c_s) \equiv 0 \pmod{p}$ and $p^{-s} f(c_s) \not\equiv 0 \pmod{p}$, then for this c_s we can proceed for the solution of the given congruence.

Procedure. Firstly take $s = 1$ and by taking $x = 0, 1, 2, ..., p - 1$ observe the values of x for which $f(x) \equiv 0 \pmod{p}$.

Let c_1 be a value of x for which $f(x) \equiv 0 \pmod{p}$ has a solution. Then try to find the solution of $f(x) \equiv 0 \pmod{p^2}$. To get a general solution $c_2 = c_1 + tp$, we shall solve the following linear congruence.

$$f'(c_1)\ t \equiv -\frac{f(c_1)}{p} \pmod{p}$$

From this, as per the previous rule, we get 0 or p number of values of t. We can now get the values of c_2 by taking any of the values of t. Similarly, by getting c_2 try to solve the congruence

$$f(x) \equiv 0 \pmod{p^3}$$

In order to get a general solution $c_3 = c_2 + tp^2$ of the above congruence we must have

$$f'(c_2)t \equiv -\frac{f(c_2)}{p^2} \pmod{p}$$

By solving this linear congruence, we get 0 or 1 or p number of values of t and we can get c_3 by taking any of the values of t.

As before we can get c_4 by taking value of c_3, c_5 by any value of c_4 and in general c_{i+1} by taking any value of c_i.

For some c_i, since the number of possible values of c_{i+1} is 0 or 1 or p, if we start with any c_1 and proceed as above the number of values of c_1 we will obtain be any multiple of 0 or 1 or p. Of course it is clear that in each case,

$c_2 \equiv c_1 \pmod{p}$, $c_3 \equiv c_2 \pmod{p^2}$, ..., $c_{s-1} \equiv c_s \pmod{p^s}$ ... and if we start with some specific c_1 and arrive at c_r, then

$$c_1 \equiv c_2 \equiv c_3 \equiv \ldots \equiv c_r \pmod{p}$$
$$c_2 \equiv c_3 \equiv \ldots \equiv c_r \pmod{p^2}$$
$$c_3 \equiv c_4 \equiv \ldots \equiv c_r \pmod{p^3}$$
$$c_4 \equiv c_5 \equiv \ldots \equiv c_r \pmod{p^4}$$
$$c_k \equiv \ldots \equiv c_r \pmod{p^k}$$
$$c_{r-1} \equiv c_r \pmod{p^{r-1}}$$

Corollary. Let c_1 be a solution of $f(x) \equiv 0 \pmod{p}$ such that $f'(c_1) \neq 0 \pmod{p}$. Then, the solution of $f(x) \equiv 0 \pmod{p^2}$ obtained from c_1 is unique.

Proof. Starting with c_1 and from the solution $t = t_1$ of

$$f'(c_1)t \equiv -\frac{f(c_1)}{p} \pmod{p}$$

Let $c_2 = c_2 + t_1 p$. Since $f'(c_1) \neq 0$, this c_2 is unique. Then, we obtain $c_3 = c_2 + t_2 p_2$ from the solution $t = t_2$ of

$$f'(c_2)\ t \equiv -\frac{f(c_2)}{p^2} \pmod{p}$$

But $c_1 \equiv c_2 \pmod{p}$ and so $f'(c_1) \equiv f'(c_2) \pmod{p}$. So, the last congruence is written as

$$f'(c_1)t \equiv \frac{-f(c_2)}{p^2} \pmod{p}$$

Since $f'(c_1) \equiv 0$, $t = t_2$ is unique and hence c_3 is also unique, Again, since

$$c_1 \equiv c_2 \equiv c_3 \equiv \ldots \equiv c_s \pmod{p}$$

i.e., since $f'(c_1) \equiv f'(c_2) \equiv \ldots \equiv f'(c_s) \pmod{p}$, $c_{s+1} = c_s + t_s p^s$ is unique. Because, since $f'(c_s) \equiv f'(c_1) \not\equiv 0$, the solution $t = t_s$ of

$$f'(c_s)t \equiv -\frac{f(c_s)}{p^s} \pmod{p}$$

is unique. It is clear that

$$\begin{aligned} c_r &\equiv c_{r-1} + t_{r-1}\,p^{r-1} \\ &\equiv c_{r-3} + t_{r-2}\,p^{r-2} + t_{r-1}\,p^{r-1} \\ &\equiv c_{r-3} + t_{r-3}\,p^{r-3} + t_{r-2}\,p^{r-2} + t_{r-1}\,p^{r-1} \\ &\equiv \ldots\ldots\ldots\ldots\ldots\ldots\ldots\ldots\ldots\ldots \\ &\equiv c_1 + t_1p + t_2p^2 + t_3p^3 + \ldots + t_{r-1}\,p^{r-1} \end{aligned}$$

Worked out examples

1. Solve $f(x) = x^3 - x - 4 \equiv 0 \pmod{7^2}$.

Here, $p = 7$. For the solution of $f(x) \equiv 0 \pmod 7$ by taking $x = 0, 1, 2, 3, 4, 5, 6$, we observe that if $c_1 = 3$, then $f(c_1) \equiv 0$ (mod 7). Note that if $x \not\equiv 3 \pmod 7$, then $f(x) \not\equiv 0 \pmod 7$.

Again, since

$$f'(x) \equiv 3x^2 - 1$$

and for the congruence

$$f'(c_1)t \equiv -\frac{f(c_1)}{p} \pmod p$$

Since $f'(c_1) = 3, c_1^2 - 1 = 3 \times 9 - 1 = 26$ and $f(c_1) = 28 = 4 \times 7$ we have

$$26t \equiv -\frac{4\times 7}{7} = -4 \pmod 7$$

In this case, since $26 \equiv -2 \pmod 7$,

$$-2t \equiv -4 \pmod 7 \text{ or } 2t \equiv 4 \pmod 7$$

So, since $2(t - 2)$ is divisible by 7 and $(2, 7) = 1$ we have $t = t_1 = 2$. Therefore, the given congruence has

$$c_2 = c_1 + t_1p = 3 + 2 \times 7 = 17$$

as one and unique solution mod 7^2 and each solution can be written as

$$x = 17 + 7^2s, \; s = 0, \pm 1, \pm 2, \ldots$$

2. Solve $f(x) = x^2 - x + 2 \equiv 0 \pmod{7^2}$. Here, $p = 7$ and one unique solution of $f(x) = x^2 - x + 2 \equiv 0 \pmod 7$ is

$$x \equiv c_1 = 4 \pmod 7$$

Again, $f'(x) = 2x - 1$, $f(c_1) = f(4) = 14$, $f'(c_1) = f'(4) = 7$

$$f'(c_1)t \equiv -\frac{f(c_1)}{p} \pmod p$$

will determine $c_2 = c_1 + tp = 4 + 7t$.

This is of the form

$$7t \equiv -2 \pmod 7 \text{ or, } 0.t = -2 \pmod 7.$$

But such a 't' is impossible. Therefore, no c_2 is obtained from c_1. Since $f(x) \equiv 0 \pmod 7$ has no solution other than c_1 and for this c_1, $f(x) \equiv 0 \pmod{7^2}$ has no solution, it follows that the given congruence has no solution.

3. Solve $f(x) = x^2 + x + 1 \equiv 0 \pmod{7^3}$. First of all we consider the solution of $f(x) = x^2 + x + 1 \equiv 0 \pmod 7$. Since $x^2 + x + 1 \equiv x^2 - 6x + 8 = (x - 2)(x - 4) \pmod 7$, $f(x) \equiv 0 \pmod 7$ has two and only two solutions: $c_1 = 2$, $c_1 = 4$.

(*i*) Take $c_1 = 2$. Since $f(c_1) = f(2) = 7$ and $f'(c_1) = f'(2) = 5$, in order to obtain $c_2 = c_1 + tp$, $f'(c_1)\, t \equiv -\dfrac{f(c_1)}{p} \pmod p$ takes the form

$5t \equiv -1 \pmod 7$

Multiplying 4 on both sides of this, we get

$-t \equiv 20t = -4 \pmod 7$ or, $t = 4$.

So, $c_2 = c_1 + tp = 2 + 4 \times 7 = 30$.

Then, since $f'(c_1) = 5 \not\equiv 0 \pmod p$ to get $c_3 = c_2 + tp^2$, since

$$f'(c_1) = f'(2) = 5$$

$$f(c_2) = f(30) = 30^2 + 30 + 1 = 931, \text{ the congruence}$$

$$f'(c_1)t \equiv -\frac{f(c_2)}{p^2} \pmod p$$

i.e., $f'(2)t \equiv -\dfrac{f(30)}{7^2}$ (mod 7) takes the form

$$5t \equiv -\frac{931}{7^2} = -19 \equiv 2 \pmod 7$$

Multiplying both sides by 3, we get

$$15t \equiv 6 \pmod 7 \text{ or, } t \equiv 6 \pmod 7.$$

So taking $t_2 = 6$, one solution of the given congruence is obtained as

$$c_3 = c_1 + c_2p + tp^2$$
$$= 2 + 4 \times 7 + 6 \times 7^2 = 324.$$

(*ii*) On the other hand, if $c_1 = 4$, for the solution $c_2 = c_1 + tp$ of $f(x) = x^2 + x + 1 \pmod{7^2}$, we take the help of

$$f'(c_1)t \equiv -\frac{f(c_1)}{p} \pmod p$$

and since $f'(4)t \equiv -\dfrac{f(4)}{7} \pmod 7$

or, $f'(4) = 2 \times 4 + 1 = 9$ and $f(9) = 4^2 + 4 + 1 = 21$, we get

$$9t \equiv -\frac{21}{7} \pmod 7$$

or $2t \equiv -3 \pmod 7$

Multiplying 4 on both sides, we get

$$t \equiv 8t \equiv -12 \equiv 2 \pmod 7$$

So, $c_2 = c_1 + tp = 4 + 2 \times 7 = 18$.

Thus with the help of

$$f'(c_1)t \equiv -\frac{f(c_2)}{p} \pmod p, \text{ we obtain}$$

$$2t \equiv 9t \equiv -\frac{343}{49} = -7 \equiv 0 \pmod 7 \text{ or, } t \equiv 0 \pmod 7$$

So, $c_3 = c_2 + t = 18 + 0 \times 7 = 18$ and hence $c_1^{\,3} = 18$ is the solution of $f(x) \equiv 0 \pmod{7^3}$. Thus, we get two solutions

$$x \equiv 324,\ 18 \pmod{7^3} \text{ of } x^2 + x + 1 \equiv 0 \pmod{7^3}.$$

EXERCISE

Solve the following congruences:

1. $x^3 - x + 4 \equiv 0 \pmod{7^3}$ **Ans:** $x \equiv 66 \pmod{7^3}$
2. $x^2 + x + 1 \equiv 0 \pmod{7^4}$ **Ans:** (*i*) $x \equiv 1553 \pmod{7^4}$
(*ii*) $x \equiv 1047 \pmod{7^4}$
3. $x^2 - x - 12 \equiv 0 \pmod{7^3}$ **Ans:** (*i*) $x \equiv 4 + 49t \pmod{7^3}$
(*ii*) $x \equiv -3 + 49t \pmod{7^3}$
$t = 0, 1, 2, 3, 4, 5, 6$ (in total 14 solutions).
4. Prove that if $f(x) = x^3 - 2x + 3$, then
(*a*) the solution of $f(x) \equiv 0 \pmod 3$ is $x \equiv 0, 2 \pmod 3$,
(*b*) the solution of $f(x) \equiv 0 \pmod 9$ is $x \equiv 6, 5 \pmod 9$,
(*c*) the solution of $f(x) \equiv 0 \pmod{27}$ is $x \equiv 6, 23 \pmod{27}$,
(*d*) the solution of $f(x) \equiv 0 \pmod{81}$ is $x \equiv 60, 23 \pmod{81}$.
5. Show that $x^3 + x + 7 \equiv 0 \left(\bmod \dfrac{4}{3}\right)$ has no solutions.
6. If $f(x) = x^3 + x^2 - 4$, then show that $f(x) \equiv 0 \pmod{7^8}$ has no solutions.
7. If $f(x) = x^2 + x + 7$, then show that
(*a*) the solutions of $f(x) \equiv 0 \pmod 7$ is $x \equiv 0, 6 \pmod 7$
(*b*) the solution of $f(x) \equiv 0 \pmod{7^2}$ is
$$x \equiv -7, 6 \pmod{7^2}$$
(*c*) the solution of $f(x) \equiv 0 \pmod{7^3}$ is
$$x \equiv -56, 55 \pmod{7^3}.$$

Further Application of Chinese Remainder Theorem

Let us first put it in the following manner. Let $m_1, m_2, m_3, ..., m_n$ be each greater than 1 and coprime. Then for any integers $c_1, c_2, c_3, ..., c_n$ the following simultaneous congruences

$$c \equiv c_1 \pmod{m_1}$$
$$c \equiv c_2 \pmod{m_2}$$
$$c \equiv c_3 \pmod{m_3}$$
$$\vdots$$
$$c \equiv c_n \pmod{m_n} \text{ has a solution.}$$

Let $m = m_1, m_2, m_3, ..., m_n$ and for each $i, j, (mi, mj) = 1$. We now discuss the solution for the following simultaneous congruences

$$f(x) \equiv c_1 \pmod{m_1}$$
$$f(x) \equiv c_2 \pmod{m_2}$$
$$\vdots$$
$$f(x) \equiv 0 \pmod{m_n}.$$

We call this a *System of Congruences*. If c is a solution of each of the above congruences, then c is said to be a solution of the system. We have to determine a c such that for $i = 1, 2, ..., n$, we must have $f(c) \equiv 0 \pmod{m_i}$.

Suppose that for each i, m_1 divides $f(c)$. Then since m_1 divides $f(c)$, $f(c) = m_1 f_1(c)$. Again, m_2 divides $m_1 f_1(c) = f(c)$ and $(m_1, m_2) = 1$. So, we can get two integers n_1, n_2 such that $m_1 n_1 + m_2 n_2 = 1$. So m_2 divides

$n_1 f(c) + m_2 n_2 f_1(c) = (m_1 n_1 + m_2 n_2) f_1(c) = f_1(c)$, and hence $f_1(c) = m_2 f_2(c)$. Similarly, $f_2(c) = m_3 f_3(c)$, $f_3(c) = m_4 f_4(c)$, ..., $f_n(c) = m_n g(c)$, *i.e.*,

$$f(c) \equiv m_1 m_2 m_3 \dots m_n g(c)$$

i.e., $$f(c) \equiv 0 \pmod{m}$$

Clearly, if $f(c) \equiv 0 \pmod{m}$, then for $i = 1, 2, ..., n$, $f(c) \equiv 0 \pmod{m_i}$. We showed that all the solutions of $f(x) \equiv 0 \pmod{m}$ are obtained from $f(x) \equiv 0 \pmod{m_i}$, $i = 1, 2, ..., n$ and hence it is enough to obtain all the solutions of this system.

Suppose that for each i, c is a solution of $f(x) \equiv 0 \pmod{m_i}$. Now consider the set $\{c_1, c_2, ..., c_n\}$ obtained in this way and obtain a c by using Chinese Remainder theorem such that

$$
\begin{aligned}
c &\equiv c_1 \pmod{m_1}\\
c &\equiv c_2 \pmod{m_2}\\
&\vdots\\
c &\equiv c_i \pmod{m_i}\\
&\vdots\\
c &\equiv c_n \pmod{m_n}
\end{aligned}
$$

Then,

$$
\begin{aligned}
f(c) &\equiv f(c_1) \equiv 0 \pmod{m_1}\\
f(c) &\equiv f(c_2) \equiv 0 \pmod{m_2}\\
&\vdots\\
f(c) &\equiv f(c_i) \equiv 0 \pmod{m_i}\\
&\vdots\\
f(c) &\equiv f(c_n) \equiv 0 \pmod{m_n}
\end{aligned}
$$

So, c is a solution of $f(x) \equiv 0 \pmod{m}$.

Now, we determine the number of distinct solutions of $f(x) \equiv 0 \pmod{m}$. Let this be $N_f(m)$. Similarly, assume that the number of distinct solutions of $f(x) \equiv 0 \pmod{m_i}$ is $N_f(m_i)$. We now prove that

$$N_f(m) = N_f(m_1)\, N_f(m_2) \,...\, N_f(m_n)$$

Let s_m be subset of $\{0, 1, 2, ..., (m-1)\}$ consisting of solutions of $f(x) \equiv 0 \pmod{m}$ and s_{mk} the subset of $\{0, 1, 2, ..., (m_k-1)\}$ consisting of solutions of $f(x) \equiv 0 \pmod{m_k}$. Let $c \in s_m$, *i.e.*, $m \mid f(c)$. Observe that for each i, $m_i \mid f(c)$. Let r_i be the remainder when c is divided by m_i. So, we obtain a set $\{r_1, r_2, ..., r_n\}$ of remainders. Again, $c \equiv r_i \pmod{m_i}$.

Hence

$$f(r_i) \equiv f(c) \equiv 0 \pmod{m_i}$$

i.e., $r_i \in s_{mi}$. Thus the ith element of the set $\{r_1, r_2, ..., r_n\}$ of remainders belongs to s_{mi}. On the other hand, if a set of remainders $\{r_1, r_2, ..., r_n\}$

is given, then by Chinese Remainder theorem we can obtain a c such that $c \in s_m$ and for each i, $c \equiv r_i \pmod{m_i}$. Again, if for some $d \in s_m$, $d \equiv r_i \pmod{m_i}$, $i = 1, 2, ..., n$, then c and d will be identical, *i.e.*, given a particular set of remainders we obtain one and only one $c \in s_m$. Conversely, for each $c \in s_m$, we obtain a definite set of remainders.

Therefore, in order to obtain all possible sets of remainders, we take any element of s_{mi} as a possible value of r_i for $i = 1, 2, ..., n$.

Again, the number of solutions of $f(x) \equiv 0 \pmod{m_1}$ which are in $\{0, 1, 2, ..., m_1-1\}$ is $N_f(m)$. Since r_1 is a remainder $r_1 \in \{0, 1, 2, .., m_1 - 1\}$ and since $f(r_1) \equiv 0 \pmod{m_1}$, r_1 is a solution of $f(r_1) \equiv 0 \pmod{m_1}$. So, the possible values of r_1 from the above set of solutions of $f(x) \equiv 0 \pmod{m_1}$. So, the number of possible distinct values of r_1 is $N_f(m_1)$, *i.e.*, the number of possible values of r_1 from s_{m1} is $N_f(m_1)$. Similarly, the possible values of r_2 from s_{m_2} is $N_f(m_2)$. So, the possible number of sets $\{r_1, r_2\}$ is $N_f(m_1)$, $N_f(m_2)$. Continuing in this way, we observe that the possible number of sets $\{r_1, r_2, ..., r_n\}$ is $N_f(m_1)$, $N_f(m_2)$ ... $N_f(m_n)$. For a specific set $\{r_1, r_2, ..., r_n\}$, $f(x) = 0 \pmod{m}$ has one and only one solution $x = c$ and therefore,

$$\begin{aligned} N_f(m) &= \text{the number of distinct solutions of } f(x) \equiv 0 \pmod{m} \\ &= \text{the number of possible sets } \{r_1, ..., r_n\} \\ &= N_f(m_1).\ N_f(m_2) \ ......\ N_f(m_n). \end{aligned}$$

Note: If $p_1, p_2, ..., p_n$ are distinct prime numbers and if $\boldsymbol{m} = p_1^{a_1} p_2^{a_2} ..., p_n^{a_n}$, then if we take $m_k = p_k^{a_k}$, the application of the above procedure becomes simpler.

Example. $x^3 + 90x^2 + 83x + 99 \equiv 0 \pmod{3^2.\ 5^2.\ 7}$

Let $f(x) = x^3 + 90x^2 + 83x + 99$

Firstly we need the solutions of each of the following congruences:

(I) $f(x) \equiv 0 \pmod{3^2}$ (II) $f(x) \equiv 0 \pmod{5^2}$

(III) $f(x) \equiv 0 \pmod{7}$

Solution of (I): If $F(x) = x^3 + 2x$, then $f(x) = F(x) \pmod{3^2}$. Here $F(x) \equiv x^3 - 3x^2 + 2x = x(x-1)(x-2) \equiv 0 \pmod{3}$. So the solutions of $F(x) \equiv 0 \pmod{3}$ are $x \equiv 0, 1, 2 \pmod{3}$. Again, since

$F(x) = 3x^3 + 2 \equiv 2 \pmod 3$, by taking $c_2 = c_1 + 3t$ in

$F(c_1).t \equiv -\dfrac{F(c_1)}{p} \pmod p$ we obtain the following values of t and c_2

when $c_1 = 0$, 1 and 2.

Value of c_1	Value of t	Value of c_2
0	0	0
1	1	4
2	1	5

So, the solutions of (I) are $x \equiv 0, 4, 5 \pmod{3^2}$.

Solution of (II): If $\phi(x) = x^3 + 15x^2 + 8x - 1$, then $f(x) \equiv \phi(x) \pmod{5^2}$. From $\phi(x) \equiv x^3 + 3x - 1 \equiv (x-3)^2 (x-4) \equiv 0 \pmod 5$ we obtain the values of x as $c_1 \equiv 3$ or $4 \pmod 5$. Since $\phi(c_1) \equiv 0 \pmod 5$, if $c \equiv 3 \pmod 5$, we cannot get the value of t by using

$\phi(c).t \equiv \dfrac{-\phi(c_1)}{p} \pmod p$. But if $c_1 \equiv 4 \pmod 5$, then we get $c_2 = c_1 + 5t$

$= c_1 + 5 \times 0 = 4$. So the solution of (*ii*) is $x \equiv 4 \pmod{5^2}$.

Solution of (III): If $F(x) = x^3 - x^2 - 1 = (-x + 1)(x - 1)^2$, then since $f(x) \equiv F(x) \pmod 7$, $x \equiv 1, 1 \pmod 7$.

So, the number of solutions of $f(x) \equiv 0 \pmod{3^2 . 5^2 . 7}$ is $3 \times 1 \times 2 = 6$ and in order to obtain these solutions, we apply Chinese Remainder theorem. For $a_1 = 0$, 4 or 5, $a_2 = 4$ and $a_3 = 1$ or -1, we obtain the solutions of $x \equiv a_1 \pmod{3^2}$, $x \equiv a_2 \pmod{5^2}$, $x \equiv a_3 \pmod 7$. So, $x \equiv 419, 594, 769, 869, 1044, 1219 \pmod{3^2 . 5^2 . 7}$.

EXERCISE

1. Solve the following congruences:

   (*a*) $x^2 + 2x - 5 \equiv 0 \pmod{15}$ **Ans:** $x \equiv 5 \pmod{15}$

   (*b*) $x^2 + x + 7 \equiv 0 \pmod{189}$

   Ans: $x \equiv 13, 49, 76, 112, 139, 175 \pmod{189}$

   (*c*) $x^3 - 4x^2 - 11x + 30 \equiv 0 \pmod{419}$

   Ans: $x \equiv 2, 5, -3 \pmod{419}$

   (*d*) $x^3 - 4x^2 - 11x + 30 \equiv 0 \pmod{463}$

   Ans: $x \equiv 2, 5, -3 \pmod{463}$

(*e*) $x^3 - 4x^2 - 11x + 30 \equiv 0 \pmod{91}$.

Ans: $x \equiv 2, 5, 23, 44, 54, 67, 75, 18, -3 \pmod{91}$.

2. Without trying to solve, find the number of solutions of the following congruences:

(*a*) $x^3 - x + 1 \equiv 0 \pmod{3^{10}.13^5}$

(*b*) $x^3 - x + 1 \equiv 0 \pmod{5^7.7^5}$

(*c*) $x^3 - x + 1 \equiv 0 \pmod{7^8.11^4}$

(*d*) $x^3 + 5x - 3 \equiv 0 \pmod{3^{10}.5^5}$

(*e*) $x^3 + 2x - 3 \equiv 0 \pmod{3^5.5^2}$.

Ans: (*a*) 0, (*b*) 1, (*c*) 1, (*d*) 3, (*e*) = 18.

2.3 Euler's Function ϕ

Let m be a given positive integer. Then, the number of positive integers less than m and coprime with m is denoted by $\phi(m)$ and ϕ is known as the Euler's function.

If a is coprime with m and $x \equiv a \pmod{m}$, then x is also coprime with m. We first prove

Theorem 8. *If* $(k, m) = 1$ *and if* $a_1, a_2, ..., a_m$ *form a complete system of incongruent residues mod m, then* $ka_1, ka_2, ..., ka_m$ *are also a complete system of incongruent residues mod m.*

Proof. If $ka_i \equiv ka_j \pmod{m}$, then $k(a_i - a_j) \equiv 0 \pmod{m}$ or, $m \mid k(a_i - a_j)$. Since $(k, m) = 1$, we can choose positive integers x and y such that $kx + my = 1$. So, m will divide $k(a_i - a_j)x + m(a_i - a_j)y = a_i - a_j$. But in order that this is possible $i = j$. So, $ka_1, ka_2, ..., ka_m$ form a complete system of incongruent residues.

Theorem 9. *Let* $(k, m) = 1$. *If* $a_1, a_2, ..., a\phi_{(m)}$ *form a complete system of incongruent residues mod m and each coprime with m, then* $ka_1, ka_2, ..., ka\phi_{(m)}$ *form a complete system of incongruent residues mod m and each coprime with m.*

Proof. As before in order that $ka_i = ka_j$, we must have $i = j$. Since each of k and $a_1, a_2, ..., a_{\phi(m)}$ is coprime with m each of $ka_1, ka_2, ..., ka_{\phi(m)}$ is coprime with m.

Theorem 10. *Let* $(m, m') = 1$. *If a belongs to a complete system of incongruent residues mod m and a belongs to a complete system of*

incongruent residues mod m', then $(a'm + m'a)$ belongs to a complete system of incongruent residues mod mm'.

Proof. a can take values 0, 1, 2, 3,... , $m - 1$. Let $a = r$. Then, by taking $a' = 0, 1, 2, ..., m' - 1$ successively, we get the following values of $(a'm + am')$: $0 . m + r . m'$, $1 . m + r . m'$, $2 . m + r.m'$, ..., $(m' - 1) m + r. m'$. Thus, for a particular value of a, we get m' number of values for $(a'm + am')$. Therefore, for all the values of a, the numbers are incongruent to each other mod (mm'), and form a complete system of incongruent residues mod (mm').

A function f is said to be *multiplicative* if $f(mm') = f(m) f(m')$ for $(m, m') = 1$.

Theorem 11. *The Euler's function ϕ is multiplicative.*

Proof. Let $(m, m') = 1$. We have to show that $\phi(mm') = \phi(m) \phi(m')$.

If a belongs to a complete system of incongruent residues mod m and a' belongs to a complete system of incongruent residues mod m', then by Theorem 10, $(a'm + am')$ belongs to a complete system of incongruent residues mod (mm'). If $(a'm + am')$ are coprime, then $(a'm + am')$ and m are coprime and also $(a'm + am')$ and m' are coprime. Hence am' and m are coprime, also $a'm$ and m' are coprime. Therefore, a' and m are coprime, and a' and m' are coprime.

Conversely, if a and m are coprime, and a' and m' are coprime, then am', m are coprime and $a'm$, m' are coprime. Consequently, $(a'm + am')$, m are coprime, $(a'm + am')$, m' are coprime. Hence $(a'm + am')$ and mm' are coprime.

Now 'a' can take $\phi(m)$ values each less than m and coprime with m. a' can take $\phi(m')$ values each less than m' and coprime with m'. Hence $(a'm + am')$ has exactly $\phi(m) \phi(m')$ different values, each less than mm' and coprime with mm'. Therefore,

$$\phi(mm') = \phi(m) \phi(m')$$

Along with Theorem 11 we have also proved.

Theorem 12. *If $(m, m') = 1$, a belongs to a complete system of incongruent residues each coprime with m, a' belongs to a complete system of incongruent residues each coprime with m', then $(a'm + am')$ belongs to a complete system of incongruent residues each coprime with (mm').*

We denote the product of $a_1, a_2, ..., a_n$ as $\pi_{k=1}^{n} a_k$ or briefly as πa.

Since each positive integer is a product of certain positive prime numbers and since ϕ is multiplicative, we proceed as follows;

Let p be a prime number and $m = p^c$. The number of positive integers less than p^c is p^c-1. Among these, the number of positive integers which are multiples of p is $(p^{c-1}-1)$, these are

$$p, 2p, 3p, 4p, ..., p^c-p$$

and others are coprime with p, Therefore,

$$\phi(p^c) = (p^c - 1) - (p^{c-1} - 1) = p^c - p^{c-1}$$

$$= p^c\left(1-\frac{1}{p}\right)$$

Corollary. If $p_1, p_2, p_3, ..., p_n$ are prime factors of m and if

$$m = p_1^{c_1} p_2^{c_2} ... p_n^{c_n}$$

then
$$\phi(m) = m\left(1-\frac{1}{p_1}\right)\left(1-\frac{1}{p_2}\right)...\left(1-\frac{1}{p_n}\right).$$

Proof. $\phi(m) = \phi(p_1^{c_1} p_2^{c_2} ... p_n^{c_n})$

$$= \phi(p_1^{c_1})\,\phi(p_2^{c_2})...\phi(p_n^{c_n})$$

$$= p_1^{c_1}\left(1-\frac{1}{p_1}\right) p_2^{c_2}\left(1-\frac{1}{p_2}\right)... p_n^{c_n}\left(1-\frac{1}{p_n}\right)$$

$$= m\left(1-\frac{1}{p_1}\right)\left(1-\frac{1}{p_2}\right)...\left(1-\frac{1}{p_n}\right).$$

In short, we write the above statement as follows:

If $m = \pi p^c$, then $\phi(m) = m\pi_{p|m}\left(1-\frac{1}{p}\right)$.

In the above statement, it is implicit that p is a prime number.

In the following theorems, we discuss some more properties of ϕ. Firstly, we prove

Theorem 13. *If p is a prime number, then* $a^p \equiv a \pmod{p}$.

Proof. If $p \mid a$, then since $a^p \equiv 0 \equiv a \pmod{p}$, the theorem holds. Assume that $p \mid a$. Now $\{1, 2, 3, 4, ..., (p-1)\}$ forms a complete system of incongruent residues mod p. Since $(p, a) = 1$, from Theorem 8, it follows that

$$\{a, 2a, 3a, ..., (p-1)a\}$$

also form a complete system of incongruent residues mod p. Consider the following sets:

$$A = \{1, 2, 3, ..., (p-1)\}$$

$$B = \{a, 2a, 3a, ..., (p-1)a\}.$$

A and B are such that one element of A is congruent to one and only one element of B mod p. So the product of elements of A is congruent to the product of elements of A mod p, *i.e.*,

$$1, 2, 3, ..., (p-1) \equiv a, 2a, 3a, ..., (p-1)a \pmod{p}$$

or $$1.2.3, ..., (p-1) \equiv a^{p-1}\, 1.2.3...., (p-1) \pmod{p}.$$

Let $K = 1.2.3. ... (p-1)$. Now K and p are coprime

and $$K\, a^{p-1} \equiv K.1 \pmod{p}$$

Now, by Theorem 4,

$$a^{p-1} \equiv 1 \pmod{p}$$

Multiplying a on the both sides, we get

$$a^p \equiv a \pmod{p}$$

We have now proved that

Theorem 14. *If p is a prime number and p does not divide a, then*

$$a^{p-1} \equiv 1 \pmod{p}$$

This theorem is named after Fermat and is known as *Fermat's little theorem.*

The following theorem is proved by adapting the arguments of Theorem 13.

Theorem 15. (The Fermat-Euler Theorem)

If a and m are coprime, then

$$a^{p(m)} \equiv 1 \pmod{m}.$$

Proof. By theorem 9, $\{x_1, x_2, ..., x_{\phi(m)}\}$ forms a complete system of incongruent residues mod m each coprime with m.

Then, $$\{ax_1, ax_2, ..., ax_{\phi(m)}\}$$

forms a complete system of incongruent residues mod m each coprime with m. So, any element of first set is congruent to one and only one element of the second set mod m. Therefore the elements of the first set is congruent mod m with the product of elements of the second set, *i.e.*,

$$\prod_{k=1}^{\phi(m)} ax_k \equiv \prod_{k=1}^{\phi(m)} x_k \pmod{m}$$

i.e.,

$$a^{\phi(m)}\left(\prod_{k=1}^{\phi(m)} x_k\right) \equiv \left(\prod_{k=1}^{\phi(m)} x_k\right) \pmod{m}$$

Each x being coprime with m, their product is also coprime with m. So, by Theorem 4

$$a^{\phi(m)} \equiv 1 \pmod{m}$$

Remark. If a and m are not coprime, the theorem does not hold. For example, take $a = 2$, $m = 10$.

Then,

$$\phi(m) \equiv 4 \text{ and } a^{\phi(m)} = 2^4 = 16 \not\equiv 1 \pmod{10}.$$

2.4. Quadratic Residues: Let p be an odd prime number and $p \mid a$. If $x \in \{1, 2, 3, ..., (p-1)\}$, then exactly one of $\{1.x, 2.x, 3.x, ..., (p-1)x\}$ is congruent to a mod p. Let this one be $x'x$. Then x' is said to be an *associate* of x. In this case,

$$x'x \equiv a \pmod{p},\ 0 < x' < p.$$

For a given integer 'a', it is clear that any one of the following two possibilities hold:

(*i*) At least one of 1, 2, 3, ..., $p-1$, say x, is equal to its associate x'

(*ii*) None of 1, 2, ..., $p-1$ is equal to its associate.

Let the possibility (*i*) hold and let x be its own associate.

Then, $$x^2 \equiv a \pmod{p}.$$

$x = x_1$ is a solution of the above congruence, In this case a is said to be *Quadratic residue* of p. It is clear that,

since

$$x_1^2 \equiv a \pmod{p},$$

we have $$(p-x_1)^2 \equiv a \pmod{p}$$

So, $(p - x_1)$ is another solution of the congruence. Again, if x_1 and x_2 are solutions of

$$x^2 \equiv a \pmod{p},$$

then, since $x_1^2 \equiv a$ and $x_2^2 \equiv a$, we have

$$(x_1 - x_2)(x_1 + x_2) = x_1^2 - x_2^2 \equiv 0 \pmod{p}.$$

So, $(x_1 - x_2)$ or $(x_1 + x_2)$ is divisible by p. If $(x_1 - x_2)$ is divisible by p, then x_1 and x_2 are congruent mod p. But x_1 and x_2 are in $\{1, 2, 3, ..., p-1\}$ and no two of these are congruent. Therefore, $x_1 = x_2$.

If $x_1 + x_2$ is divisible by p, then since both of x_1 and x_2 are smaller than p,

$$x_1 + x_2 = p$$

because $0 < x_1$, $0 < x_2$, $x_1 + x_2 < (2p)$. So, $x_2 = p - x_1$. Thus if $x_2 \equiv a \pmod{p}$, $0 < x < p$, then x will be equal to x_1 or $p - x_1$, *i.e.*, among

$$1, 2, 3, ..., p-1$$

x_1 and $(p - x_1)$ are associates and rest $(p - 3)$ numbers will be divided into ½ $(p - 3)$ pairs such that in each pair, any one is an associate of the other. In other words, if x is an associate of x', then x' is an associate of x and they form a pair (x, x'). Rest $(p - 3)$ numbers form $\frac{1}{2}(p - 3)$ pairs in this manner.

(*ii*) If one of 1, 2, 3, ..., $(p - 1)$ is equal to its associate, then a is said to be a *Quadratic non-residue* of p. In this case, 1, 2, 3, ..., $(p - 1)$ will be divided into $\frac{1}{2}(p - 1)$ unequal pairs such that in each pair, one is an associate of other.

In the first case, if a is quadratic residue of p, then we can get an integer x_1 such that

$$x_1(p - x_1) \equiv -x_1^2 \equiv -a \pmod{p}$$

Again, for a pair of associates x, x',

$$x\,x' \equiv a \pmod{p}$$

Hence,

$$1.2.3. \ldots, (p - 1) = (p - 1)!$$

$$x_1(p - x_1)\ \pi\ xx' = -a.a^{\frac{1}{2}(p - 3)} \pmod{p}$$

Clearly for any p,

$$1^2 \equiv 1 \pmod{p}$$

$$(p - 1)^2 \equiv 1 \pmod{p}$$

i.e., for any p, 1 is a quadratic residue. Hence, by taking $a = 1$, we get

Theorem 16. (Wilson's Theorem)

$$(p - 1)\,! \equiv 1 \pmod{p}$$

Secondly, if a has no quadratic residue, then

$$(p - 1)\,! = \pi\ xx' = a^{\frac{1}{2}(p - 1)} \pmod{p}$$

We now introduce the definition of *Legendre's Symbol* $\left(\dfrac{a}{p}\right)$. Let p be an odd prime number; p does not divide a.

Then,

$$\left(\frac{a}{p}\right) \equiv \begin{cases} +1, \text{ if } a \text{ is a quadratic residue of } p, \\ -1, \text{ if } a \text{ is not a quadratic residue of } p. \end{cases}$$

It is clear that, if

$$a \equiv b \pmod{p}$$

then, $$\left(\frac{a}{p}\right) = \left(\frac{b}{p}\right).$$

We have just now proved that if a is a quadratic residue of p; then

$$p - 1 ! \equiv - a^{½(p-1)} \pmod{p}$$

and if a is not a quadratic residue of p, then

$$(p - 1) ! \equiv a^{½(p-1)} \pmod{p}$$

By using Legendre's symbol, it can be restated as:

Theorem 17. *If p is an odd prime number and p does not divide a, then*

$$(p - 1) ! \equiv -\left(\frac{a}{p}\right) a^{1/2(p-1)} \pmod{p}$$

By using Wilson's theorem, this can be written as:

$$-1 \equiv (p - 1) ! \equiv -\left(\frac{a}{p}\right) a^{1/2(p-1)} \pmod{p}$$

By multiplying $-\left(\frac{a}{p}\right)$ on both sides, we obtain the following theorem.

Theorem 18.

$$\left(\frac{a}{p}\right) \equiv a^{½(p-1)} \pmod{p}$$

2.5. Some Properties of Quadratic Residues and Quadratic Non-Residues

For a given odd number p, first of all, we will determine the number of quadratic residues and non-residues. We claim that

$$1^2, 2^2, 3^2, ..., \{½(p - 1)\}^2$$

are incongruent to each other. For otherwise, if $r^2 \equiv s^2 \pmod p$, then $(r - s)(r + s)$ is divisible by p. Since $0 < r + s < p - 1$, $(r + s)$ is not divisible by p, so p must divide $(r - s)$. But both r and s are smaller than p. So in order that $p \mid (r - s)$, we must have $r = s$. That is, if $0 < r, s < ½(p - 1)$ and if

$$r^2 \equiv s^2 \pmod p, \text{ then } r = s.$$

Thus, our claim is established

and $$1^2, 2^2, 3^2, ..., \{½(p - 1)\}^2$$

are incongruent mod p, Again, these are respectively congruent to

$$(p - 1)^2, (p - 2)^2, (p - 3)^2, ..., \{½(p + 1)\}^2$$

because

$$r^2 \equiv (p - r)^2 \pmod p$$

If we take $r^2 = a$, then we observe that

$$x^2 \equiv a \pmod p$$

$x = p - r$ is a solution of this congruence. So, $a = r^2$ is a quadratic residue mod p. But r can take any value from among 1, 2, 3, ..., $½(p - 1)$. Thus, each odd prime p has at least $½(p - 1)$ quadratic residues.

Theorem 19.

(*i*) *The product of two quadratic residues is a quadratic residue.*

(*ii*) *The product of two quadratic non-residues is a quadratic residue.*

(*iii*) *The product of one quadratic residue with a quadratic non-residue is a quadratic non-residue.*

Proof. (*i*) Let $\alpha_1, \alpha_2, ...,$ be quadratic residues, Then, $x = x_1$ and $y = y_1$ are respectively the solutions of

$$x^2 \equiv \alpha_1 \pmod{p}$$

and $$y^2 \equiv \alpha_1 \pmod{p}.$$

Then, $$Z \equiv x_1 y_1 \text{ is a solution of}$$

$$Z \equiv \alpha_1\alpha_2 \pmod{p}$$

and thus, $\alpha_1\alpha_2$ is a quadratic residue.

(*ii*) If β_1 and β_2 are two quadratic non-residues, then

$$\beta_1^{1/2(p-1)} \equiv \left(\frac{\beta_1}{p}\right) = -1 \pmod{p}$$

$$\beta_2^{1/2(p-1)} \equiv \left(\frac{\beta_2}{p}\right) = -1 \pmod{p}$$

So, $$(\beta_1\beta_2)^{1/2(p-1)} \equiv +1 \pmod{p}$$

But $$(\beta_1\beta_2)^{1/2(p-1)} \equiv \left(\frac{\beta_1\beta_2}{p}\right)$$

So, $$\left(\frac{\beta_1\,\beta_2}{p}\right) \equiv +1$$

and hence, $\beta_1\beta_2$is a quadratic residue.

(*iii*) If β is a quadratic nonresidue, then

$$\left(\frac{\beta}{p}\right) \equiv \beta^{1/2(p-1)} \pmod{p}$$

i.e., $$\beta^{1/2(p-1)} \equiv -1 \pmod{p}.$$

If α is a quadratic residue, then

$$\left(\frac{\alpha}{p}\right) \equiv \alpha^{1/2(p-1)} \pmod{p}$$

or $$\alpha^{1/2(p-1)} \equiv +1 \pmod{p}$$

So, $$(\alpha\beta)^{1/2(p-1)} \equiv -1 \pmod{p}$$

But $$(\alpha\beta)^{1/2(p-1)} \equiv \left(\frac{\alpha\beta}{p}\right)$$

So, $$\left(\frac{\alpha\beta}{p}\right) \equiv -1, \text{ i.e., } \alpha\beta \text{ is a quadratic non-residue.}$$

It is now natural to ask: whether for each odd prime p, there is atleast one quadratic non-residue. Firstly, observe that a number which is a quadratic residue mod p is congruent to one of $1^2, 2^2, 3^2, \ldots, \{\frac{1}{2}(p-1)^2\}$.

For any integer x, it can be written as:

$$x \equiv pq + r, \; 0 < r < p$$

So, $$x^2 \equiv r^2, \; 0 < r < p$$

So each of the numbers $a_1, a_2, \ldots$ which are quadratic residues is congruent to one of,

$$1^2, 2^2, 3^2, \ldots, (p-1)^2$$

Again, mod p, $$\left\{\frac{1}{2}(p+1)\right\}^2, \left\{\frac{1}{2}(p+3)\right\}^2, \ldots (p-1)^2$$

are respectively congruent to

$$\left\{\frac{1}{2}(p-1)\right\}^2, \left\{\frac{1}{2}(p-3)\right\}^2, \ldots 3^2, 2^2, 1^2$$

Therefore, each of the quadratic residues $a_1, a_2, \ldots$ is congruent to one and only one of

$$1^2, 2^2, 3^2, \ldots, \left\{\frac{1}{2}(p-1)\right\}^2$$

Thus, we have proved that the number of quadratic residues mod p which are incongruent to each other is $\frac{1}{2}(p-1)$. But any complete system of incongruent residues has $(p-1)$ distinct elements. Therefore, there are $\frac{1}{2}(p-1)$ quadratic residues and $\frac{1}{2}(p-1)$ quadratic non-residues.

Theorem 20. *For every odd prime number p, the number of quadratic residues incongruent to each other is* $\frac{1}{2}(p-1)$ *and the number of such quadratic non-residues is* $\frac{1}{2}(p-1)$.

We have proved earlier that if p is an odd prime number, then

$$\left(\frac{a}{p}\right) \equiv a^{1/2(p-1)} \pmod{p}$$

If we take $a = p - 1 \equiv -1$, then we have

$$\left(\frac{p-1}{p}\right) \equiv (-1)^{1/2(p-1)}$$

Again, each such p can be expressed either as $4k + 1$ or $4k + 3$.

Theorem 21. $(p - 1)$ *is a quadratic residue for each prime p which can be expressed as 4k + 1 and (p − 1) is a quadratic non-residue for each prime p which can be expressed as 4k + 3.*

Proof. In the first case,

$$\left(\frac{p-1}{p}\right) \equiv (-1)^{1/2(p-1)} = (-1)^{2k} = 1 \pmod{p}$$

In the second case,

$$\left(\frac{p-1}{p}\right) \equiv (-1)^{1/2(p-1)} = (-1)^{2k+1} = 1 \pmod{p}.$$

Theorem 22. *Let p be a prime number which can be expressed in the form 4k + 1. Then, there is a positive integer m less than p such that (mp − 1) is a square.*

Proof. In this case, $(p - 1)$ is a quadratic residue.

So, $\quad x^2 = (p - 1) \equiv -1 \pmod{p}$

has a solution and this solution belongs to the set

$$\left\{1, 2, 3, \ldots, \frac{1}{2}(p-1)\right\}.$$

So, $$0<1+x^2<1+\left(\frac{1}{2}p\right)^2<p^2$$

and p divides $1 + x^2$. Hence there is one m such that $0 < m < p$ and $1 + x^2 = mp$. Therefore, $(mp - 1)$ is a square for this m.

Theorem 23. *For each odd prime number p, there exist x, y and m such that*

$$1 + x^2 + y^2 = mp \text{ and } 0 < m < p.$$

Proof. Any two elements of the set

$$\left\{0, 1^2, 2^2, 3^2, ..., \left\{\frac{1}{2}(p-1)\right\}^2\right\}$$

are incongruent mod p. Also, any two elements of,

$$\left\{p^2-1, p^2-1-1^2, p^2-1-2^2, ..., p^2-1-\left\{\frac{1}{2}(p-1)\right\}^2\right\}$$

are incongruent mod p. Each of the above sets has $\frac{1}{2}(p+1)$ elements and hence the union has $(p + 1)$ elements. But any complete system of incongruent residues mod p has p elements. Therefore, some element of the first set here must be congruent to some element of the second set mod p. Hence, there are integers x and y belonging to $\left\{0, 1, 2, 3, ..., \frac{1}{2}(p-1)\right\}$ for which

$$x^2 \equiv p^2 - 1 - y^2 \pmod{p}$$

i.e., p divides $(1 + x^2 + y^2)$. But for this x and y,

$$0<1+x^2+y^2<1+\left(\frac{1}{2}p\right)^2+\left(\frac{1}{2}p\right)^2<p^2$$

Therefore, for some positive integer $m < p$

$$1 + x^2 + y^2 = mp$$

We now prove the famous theorem due to Lagrange.

Theorem 24. (Lagranges Theorem). *Every positive integer is a sum of four squares.*

We first prove Theorem 24 by using Theorem 25 and then prove Theorem 25.

Theorem 25. *Every prime number is a sum of four squares.*

Proof of Theorem 24. We claim that if each of p and q can be expressed as a sum of four squares, then their product pq can also be expressed as a sum of four squares. For, let

$$p = x_1^2 + x_2^2 + x_3^2 + x_4^2,$$
$$q = y_1^2 + y_2^2 + y_3^2 + y_4^2,$$
$$z_1 = x_1y_1 + x_2y_2 + x_3y_3 + x_4y_4$$
$$z_2 = x_1y_2 - x_2y_1 + x_3y_4 - x_4y_3$$
$$z_3 = x_1y_3 - x_3y_1 + x_4y_2 - x_2y_4$$
$$z_4 = x_1y_4 - x_4y_1 + x_2y_3 - x_3y_2$$

Then, $$pq = z_1^2 + z_2^2 + z_3^2 + z_4^2$$

Since every positive integer can be expressed as a product of primes, the result follows from Theorem 25.

Remark. Such an expression may not be unique for a given positive integer. For example,

$$2^2 + 2^2 + 1^2 + 1^2 = 10 = 3^2 + 1^2 + 0^2 + 0^2.$$

Proof of Theorem 25.

Since $$2 = 1^2 + 1^2 + 0^2 + 0^2,$$

the result holds for 2. So we have to prove that the theorem holds for any prime number different from 2. Let p be an odd prime number. It now follows from Theorem 23 that

$$mp = x_1^2 + x_2^2 + x_3^2 + x_4^2 \text{ and } 0 < m < p.$$

Thus, $$mp < p^2.$$

Let $m_0 p$ be the smallest of all positive integers less than p^2 which can be expressed as a multiple of p and as a sum of four squares. If $m_0 = 1$, then the theorem is proved. In what follows we will show that m_0 cannot be greater than 1.

Let $m_0 > 1$. We first show that m_0 cannot be an even integer. If m_0 is even, then

$$x_1^{\,2} + x_2^{\,2} + x_3^{\,2} + x_4^{\,2} \text{ is even. So,}$$

$(x_1 + x_2 + x_3 + x_4)^2 = x_1^{\,2} + x_2^{\,2} + x_3^{\,2} + x_4^{\,2} + 2x_1x_2 + \ldots + 2x_3x_4$ is even and consequently, $x_1 + x_2 + x_3 + x_4$ is even. Thus, any one of the following possibilities are true:

(*a*) Each of x_1, x_2, x_3, x_4 is even.

(*b*) Each of x_1, x_2, x_3, x_4 is odd.

(*c*) Any two say, x_1 and x_2 are even and other two, x_3 and x_4 are odd.

In each case each of

$$x_1 + x_2,\ x_1 - x_2,\ x_3 + x_4,\ x_3 - x_4$$

is even. So

$$\frac{(x_1 + x_2)^2}{2} + \frac{(x_1 - x_2)^2}{2} + \frac{(x_3 + x_4)^2}{2} + \frac{(x_3 - x_4)^2}{2} = \frac{1}{2} m_0 p$$

is a positive integer and a sum of four squares. In this case, $0 < \frac{1}{2} m_0 p < p^2$. But from the definition of m_0p, it now follows that m_0p cannot be even. Thus if $m_0 > 1$, then m_0 is odd and its value is greater than or equal to 3.

If m_0 divides each of x_1, x_2, x_3, x_4, then m_0^2 will divide m_0p and consequently, m_0 divides p. But p being prime and $3 \le m_0 < p$, m_0 does not divide p. So, at least one of x_1, x_2, x_3, x_4 is not divisible by m_0. Hence, we can choose b_1, b_2, b_3, b_4 such tht for $k = 1, 2, 3, 4$,

$y_k = x_k - b_k m_0, |\, y_k \,| < \frac{1}{2} m_0$ and the value of atleast one y_k is different from zero. Thus,

$$y_1^2 + y_2^2 + y_3^2 + y_4^2 > 0.$$

Again since m_0 divides $(x_1^{\,2} + x_2^{\,2} + x_3^{\,2} + x_4^{\,2})$, m_0 will divide

$$\begin{aligned} y_1^{\,2} + y_2^{\,2} + y_3^{\,2} + y_4^{\,2} &= x_1^{\,2} + x_2^{\,2} + x_3^{\,2} + x_4^{\,2} + m_0^{\,2}(b_1^{\,2} + b_2^{\,2} + b_3^{\,2} \\ &\quad + b_4^{\,2}) - 2m_0\,(b_1x_1 + b_2x_2 + b_3x_3 + b_4x_4) \\ &= m_0[p + m_0\,(b_1^{\,2} + b_2^{\,2} + b_3^{\,2} + b_4^{\,2}) - 2\,(b_1x_1 + b_2x_2 \\ &\quad + b_3x_3 + b_4x_4)\,]. \end{aligned}$$

Again, $y_1^2 + y_2^2 + y_3^2 + y_4^2 < 4\left(\frac{1}{2}m_0\right)^2 = m_0{}^2$

Therefore,

$$y_1^2 + y_2^2 + y_3^2 + y_4^2 = m_0 m_1, \qquad (0 < m_1 < m_0)$$

Now, defining z_1, z_2, z_3, z_4 as in the proof of Theorem 24, we observe that

$$\begin{aligned} m_0^2 m_1 p &= (x_1^2 + x_2^2 + x_3^2 + x_4^2)\,(y_1^2 + y_2^2 + y_3^2 + y_4^2) \\ &= z_1^2 + z_2^2 + z_3^2 + z_4^2. \end{aligned}$$

Since
$$\begin{aligned} z_1 &= \Sigma x_k y_k = \Sigma x_k\,(x_k - b_k\, m_0) \\ &= x_1{}^2 + x_2{}^2 + x_3{}^2 + x_4{}^2 - m_0\,(b_1 x_1 + b_2 x_2 + b_3 x_3 + b_4 x_4), \end{aligned}$$

it follows that m_0 divides z_1. Similarly m_0 divides each of z_2, z_3, z_4. So, if we put

$zk = m_0 tk$, then each of t_1, t_2, t_3, t_4 is an integer.

Hence, by putting

$$z_k = m_0 t_k \text{ in}$$

$$z_1{}^2 + z_2{}^2 + z_3{}^2 + z_4{}^2 = m_0{}^2 m_1 p$$

we get

$$t_1{}^2 + t_2{}^2 + t_3{}^2 + t_4{}^2 = m_1 p.$$

Since $0 < m_1 < m_0$, this violates the definition of m_0. So, m_0 cannot be different from 1.

Thus, $$m_0 = 1$$

and $$p = x_1{}^2 + x_2{}^2 + x_3{}^2 + x_4{}^2.$$

2.6. In this section, we establish some useful facts about Legendre's symbol $\left(\frac{a}{p}\right)$.

Gauss's Lemma. Let p be a prime number and $(a, p) = 1$. Let m be an integer and $1 \le m \le \frac{1}{2}(p-1)$. Divide ma by p. If the remainder is less than $\frac{1}{2}p$, let it be denoted by α and if it is greater than $\frac{1}{2}p$, let it be denoted by β. Write $\beta = p - \beta'$, so that $\beta' < \frac{1}{2}p$.

By dividing

$$a, 2a, 3a, ..., \frac{1}{2}(p-1)a \quad ...(A)$$

by p in the above manner, let the remainders be

$$\alpha_1, \alpha_2, \alpha_3, ..., \alpha_\nu \quad ...(B)$$

and $$\beta_1, \beta_2, \beta_3, ..., \beta_\mu \quad ...(C)$$

or, $$p - \beta_1', p - \beta_2', p - \beta_3', ..., p - \beta_\mu' \quad ...(C')$$

In this case, each a_k and $\beta_i's$ are positive and less than $\frac{1}{2}p$.

If $\alpha_h = \alpha_k$, then for some m_h and m_k,

$$m_k a \equiv \alpha_k = \alpha_h \equiv m_h a \pmod p$$

or, $$(m_k - m_h)\, a \equiv 0 \pmod p$$

i.e., $$m_k - m_h \equiv 0 \pmod p$$

But since $1 \le m_h \le \frac{1}{2}(p-1)$ and $1 \le m_k \le \frac{1}{2}(p-1)$, this is impossible.

So $$\alpha_k \ne \alpha_h \text{ for } h \ne k$$

Similarly, $$\beta_i \ne \beta_j,\ i.e.,\ \beta_i' \ne \beta_j' \text{ for } i \ne j$$

Again, if $\alpha_r = \beta_j'$, then for some m_j and m_r,

$$m_r a \equiv \alpha r \pmod p \text{ and } m_j a = p - \beta_j' \pmod p$$

i.e., $(m_r + m_j)\, a$ is divisible by p. But since $0 < m_r + m_j < p$ and p does not divide a, this is impossible. So, no α_r is equal to no β_j.

So, $\alpha_1, \alpha_2, ..., \alpha_\nu$ and $\beta_1', \beta_2', ..., \beta_\mu'$ are $\frac{1}{2}(p-1)$ in number, each is positive and different from the other and none is greater than $\frac{1}{2}(p-1)$.

Hence the two sets

$$\{\alpha_1, \alpha_2, ..., \alpha_v, \beta_1, \beta_2, ..., \beta_\mu\}$$

and $\left\{1, 2, 3, ..., \frac{1}{2}(p-1)\right\}$ are equal. Therefore, for mod p,

$$a.2a.3a. ...\frac{1}{2}(p-1)\,a \equiv 1.2.3. ...\frac{1}{2}(p-1)a^{\frac{1}{2}(p-1)}$$
$$\equiv \alpha_1, \alpha_2, ..., \alpha_v\,(p-\beta'_1)\,(p-\beta'_2)\,...\,(p-\beta'_\mu)$$

i.e., $1.2.3....\,\frac{1}{2}(p-1)\,a^{1/2(p-1)} \equiv (-1)^\mu\,\alpha_1, \alpha_2, ..., \alpha_v\,\beta'_1, \beta'_2, ..., \beta'_\mu$

$$\equiv (-1)^\mu\,1.2.3. ...\,\frac{1}{2}(p-1)\,(\text{mod } p)$$

Thus since $1.2. ...\,\frac{1}{2}(p-1)$ is not divisible by p.

$$a^{1/2(p-1)} \equiv (-1)^\mu\,(\text{mod } p)$$

But (the Legendre's symbol)

$$\left(\frac{a}{p}\right) \equiv a^{1/2(p-1)}\,(\text{mod } p).$$

So for $m_1, m_2, ..., m_\mu$ if $m_{ka} \equiv \beta_k$ (mod p) and $\beta_k > \frac{1}{2}p$, then

$$\left(\frac{a}{p}\right) \equiv (-1)^\mu\,(\text{mod } p).$$

But the value of each is ±1 and hence

$$\left(\frac{a}{p}\right) \equiv (-1)^\mu$$

This relation is known as Gauss's Lemma.

Let p and q be two *odd* prime numbers. Consider the set

$$\{q, 2q, 3q, ..., \frac{1}{2}(p-1)\,q\}.$$

Let μ elements of the above set when divided by p have remainder greater that $\frac{1}{2}p$ and the rest when divided by p have remainder less than $\frac{1}{2}p$.

Let v elements from

$$\{p, 2p, 3p, \ldots \frac{1}{2}(q-1)p\}$$

when divided by q have remainder greater than $\frac{1}{2}q$ and rest have remainder less than $\frac{1}{2}q$. By the above rule,

$$\left(\frac{q}{p}\right) = (-1)^{\mu} \text{ and } \left(\frac{p}{q}\right) = (+1)^{\nu}$$

Since p and q are odd, $\frac{1}{2}(pq-1)$ is an integer.

Now, the set

$\{1, 2, 3, \ldots \frac{1}{2}(pq-1)\}$ can be classified into eight sets, and let us denote them by S_1, S_2, S_3, S_4, S_5, S_6, S_7 and S_8. They are as follows:

	When divided by p	*When divided by q*	*Number*
S_1	$1 \leq$ remainder $\leq \frac{1}{2}(p-1)$	$1 \leq$ remainder $\leq \frac{1}{2}(q-1)$	α (say)
S_2	$1 \leq$ remainder $\leq \frac{1}{2}(p-1)$	remainder $> \frac{1}{2}(q-1)$	β (say)
S_3	remainder $> \frac{1}{2}(p-1)$	$1 \leq$ remainder $\leq \frac{1}{2}(q-1)$	γ (say)
S_4	remainder $> \frac{1}{2}(p-1)$	remainder $> \frac{1}{2}(q-1)$	δ (say)

Table contd.

	When divided by p	*When divided by q*	*Number*
S_5	remainder $> \frac{1}{2}(p-1)$	remainder $= 0$	μ
S_6	$1 \le$ remainder $\le \frac{1}{2}(p-1)$	remainder $= 0$	$\frac{1}{2}(p-1)-\mu$
S_7	remainder $= 0$	remainder $> \frac{1}{2}(q-1)$	ν
S_8	remainder $= 0$	$1 \le$ remainder $\le \frac{1}{2}(q-1)$	$\frac{q-1}{2} - \nu$

A number, when divided by q, having remainder greater than $\frac{1}{2}(q-1)$ is in $S_2 \cup S_4 \cup S_7$. If this remainder is ρ, then such numbers are

$$\rho, q+\rho, 2q+\rho, \ldots, \frac{p-3}{2}q+\rho < \frac{pq-1}{2}$$

because the next possible number $\frac{p-1}{2}q+\rho$, being greater than

$\frac{p-1}{2}q+\frac{q-1}{2}=\frac{pq-1}{2}$, does not belong to the above class. Now,

by taking $\rho = \frac{q+1}{2}, \frac{q+3}{2}, \ldots, q+1$ successively, we observe that the number of elements of $S_2 \cup S_4 \cup S_7$ is

$$\frac{p-1}{2}.\frac{q-1}{2}.$$

But this being $\beta + \delta + \gamma$, we have

$$\beta + \delta + \gamma = \frac{p-1}{2}.\frac{q-1}{2}$$

Now by changing the rule of p and q one can see that the number of elements of $S_3 \cup S_4 \cup S_5$ is

$$\gamma + \delta + \mu = \frac{p-1}{2}.\frac{q-1}{2}$$

Let $1 \le a \le \frac{pq-1}{2}$ and a $\in$ S_3. Then,

$$\frac{pq+1}{2} \le pq - a \le pq - 1$$

and when $(pq - a)$ is divided by p, the remainder will be greater than $\frac{1}{2}(p - 1)$, but when divided by q, the remainder will be greater than $\frac{1}{2}(q - 1)$.

One the other hand, if $(pq - a)$ possesses this property, then $a \in S_3$. Therefore,

γ = the number of those elements of the set

$\left\{\frac{pq+1}{2}, \frac{pq+3}{2}, ..., pq-1\right\}$ which when divided by p, the remainder is not greater than $\frac{1}{2}(p - 1)$ and when divided by q, the remainder is greater than $\frac{1}{2}(q - 1)$.

Therefore,

$\beta + \gamma$ = the number of those elements of $\{1, 2, ... \frac{pq-1}{2}, \frac{pq+1}{2}, ..., pq-1\}$ which when divided by p, remainder is not greater than $\frac{1}{2}(p-1)$ and when divided by q, the remainder is greater than $\frac{1}{2}(q-1)$.

If any element a_k of the above set when divided by p has remainder x_k and when divided by q has remainder y_k and if $x_k \le \frac{1}{2}(p-1)$, $y_k > \frac{1}{2}(q-1)$, then given x_k we have $\frac{1}{2}(q-1)$ values of y_k. But x_k can take any value from

$$\left\{1, 2, 3, \ldots \frac{1}{2}(p-1)\right\}$$

Thus, we prove that

$$\beta + \gamma = \frac{p-1}{2} \cdot \frac{q-1}{2} \qquad \ldots(1)$$

We have already proved that

$$\beta + \delta + \gamma = \frac{p-1}{2} \cdot \frac{q-1}{2} \qquad \ldots(2)$$

$$\gamma + \delta + \mu = \frac{p-1}{2} \cdot \frac{q-1}{2} \qquad \ldots(3)$$

Adding (2) and (3) and subtracting (1) from this, we get

$$2\delta + \mu + \gamma = \frac{p-1}{2} \cdot \frac{q-1}{2}$$

So, $$\left(\frac{p}{q}\right)\left(\frac{q}{p}\right) = (-1)^{\mu+\gamma} = (-1)^{\frac{p-1}{2} \cdot \frac{q-1}{2}}$$

This is known as the *Gauss's law of quadratic reciprocity.*

Application (1): Take $m = 2$ in the above discussion. Then, γ is the number of integers among 2, 4, 6, ..., $p - 1$ which when divided by p, the remainder becomes less than $\frac{1}{2}p$. If $r \leq \Theta < r + 1$ (where r is an integer), then $[\Theta] = r$, where $[\Theta]$ denotes the integral part of Θ, such as: $\left[\frac{5}{2}\right] = 2, \left[\frac{1}{2}\right] = 0, \left[-\frac{3}{2}\right] = 2$ and so on. Note that if 2, 4, 6, ..., $2\left[\frac{1}{4}p\right]$ are divided by p, then the remainder is less than $\frac{1}{2}p$ and $\gamma = \left[\frac{1}{4}p\right]$ when divided by p, the others have remainder greater than p and

$$\mu = \frac{1}{2}(p-1) - \gamma = \frac{1}{2}(p-1) - \left[\frac{1}{4}p\right]$$

If $p = 4n + 1$, then $\mu = \frac{1}{2}(4n+2) - n = n+1 = \left[\frac{1}{4}(p+1)\right]$

So, $$\left(\frac{2}{p}\right) = (-1)^{\left[\frac{1}{4}(p+1)\right]}$$

If $p = 8n \pm 1$, then $\mu\left(\frac{2}{p}\right) = 1$, but if,

$p = 8n \pm 3$, then $\mu\left(\frac{2}{p}\right) = -1$. So, 2 is a quadratic residue for prime numbers of the form $(8n \pm 1)$ and is a quadratic non-residue for prime numbers of the form $(8n \pm 3)$.

Application (2): Take $m = -3$ and $p < 3$

If $1 \le a < \frac{1}{6}p$ then $\frac{1}{2}p < p - 3a < p$

If $\frac{1}{6}p < a < \frac{1}{3}p$ then $0 < p - 3a < \frac{1}{2}p$

If $\frac{1}{3}p < a < \frac{1}{2}p$ then $\frac{1}{2}p < 2p - 3a < p$

So it will be sufficient to take the number among $-3, -6, -9, \ldots, -\frac{3}{2}(p-1)$, or, among $2p - 3, 2p - 6, 2p - 9, \ldots, 2p - \frac{3}{2}(p-1)$ which when divided by p the remainder lies between $\frac{1}{2}p$ and p, to be $(2p - 3a)$ for

$$a = 1, 3, \ldots, \left[\frac{1}{6}p\right], \left[\frac{1}{3}p\right]+1, \left[\frac{1}{3}p\right]+2, \ldots, \left[\frac{1}{2}p\right]$$

So $\mu = \left[\frac{1}{6}p\right] + \left(\left[\frac{1}{2}p\right] - \left[\frac{1}{3}p\right]\right)$.

Hence if $p = 6n + 1$, then $\mu = n + 3n - 2n = 2n$ and if $p = 6n + 5$, then $\mu = n + (3n + 2) - (2n + 1) = 2n + 1$

So, if $p = 6n + 1$, then

$$\left(\frac{-3}{p}\right) = (-1)^{\mu} = (-1)^{2n} = 1,$$

i.e., -3 is a quadratic non-residue for the prime numbers of the form $(6n + 1)$. If $p = 6n + 5$, then

$$\left(\frac{-3}{p}\right) = (-1)^{\mu} = (-1)^{2n+1} = -1,$$

i.e., -3 is a quadratic non-residue for the prime numbers of the form $(6n + 5)$.

Application (3): If $p = 10n + k$, $k = 1, 3, 7$ or 9, then

$$\left(\frac{5}{p}\right) = \left(\frac{p}{5}\right) = \left(\frac{10n+k}{5}\right) = \left(\frac{k}{5}\right).$$

Again, since $\left(\frac{k}{5}\right) \equiv k^2 \pmod 5$, $k = 1, 4$ is a quadratic residue and $k = 2, 3$ is a quadratic non-residue for mod 5. So 5 is a quadratic residue for the prime numbers of the form $5n + 1$, $5n + 4$, $10n + 1$ and $10n + 9$ and 5 is a quadratic non-residue for the prime numbers of the form $5n + 2$, $5n + 3$, $10n + 3$, $10n + 7$. So. 5 is a quadratic residue for primes of the form $(10n \pm 1)$ and is a quadratic non-residue for primes of the form $(10n \pm 3)$.

2.7. Let a and b be integers. Then, there is some integer r for which

$b = \left[\frac{b}{a}\right]$. $a + r$, $0 \le r < a$. Let p be an odd prime, $(b, a) = 1$.

Let the number of elements of the set

$$S = \{a, 2a, 3a, \ldots \frac{p-1}{2}a\}$$

for which the smallest positive residue (mod p) is greater than $\frac{p}{2}$ be n. If we write

$$t = \left[\frac{a}{p}\right] + \left[\frac{2a}{p}\right] + \left[\frac{3a}{p}\right] + \ldots + \left[\frac{\frac{p-1}{2}a}{p}\right]$$

then $\quad (a-1)\left(\frac{p^2-1}{8}\right) \equiv (t - n) \pmod 2$

Proof. Assume that from among the positive remainders mod, p of elements of S only $r_1, r_2, \ldots, r_n$ are greater than $\frac{p}{2}$ and less than p, *i.e.*, $\frac{p}{2} < r_k < p$, $k = 1, 2, \ldots, n$. If the positive remainders smaller than $\frac{p}{2}$ are $s_1, s_2, \ldots, s_m$, then clearly for any r_k and s_i, $r_k \neq s_i$ $(k \neq i)$, we can choose r_k and s_i such that $(p - r_k) = (p - k)$. a and $s_i \equiv i.a. \pmod p$. If for some k and i, $p - r_k = s_i$, the $(i + k)\, a \equiv 0 \pmod p$ But because of $0 < i \neq k \leq \frac{p-1}{2}$ and $(a, p) = 1$, this is quite impossible. So no $(p - r_k)$ is equal to no s_i. Again more of $(p - r_k)$ or s_i is not greater than $\frac{p-1}{2}$ and they are $\frac{p-1}{2}$ in number. So the sets

$$(p - r_1, p - r_2, \ldots, p - r_n, s_1, s_2, \ldots, s_m)$$

and $$\left\{1, 2, 3, \ldots, \frac{p-1}{2}\right\}$$

are equal *i.e.*,

$$np - (r_1 + r_2 + \ldots + r_n) + (s_1 + s_2 + \ldots + s_m)$$

$$= (p - r_1) + (p - r_2) + \ldots + (p - r_n) + (s_1 + s_2 + \ldots + s_m)$$

$$= 1 + 2 + 3 + \ldots + \frac{p-1}{2} = \frac{p^2 - 1}{8}$$

Again, for $0 < l < \frac{p-1}{2}$, $la \equiv r_k$ or s_i, So

$$a.\frac{p^2-1}{8} = a + 2a + 3a + \ldots + \left(\frac{p-1}{2}\right)a$$

$$= \left\{\left[\frac{a}{p}\right] + \left[\frac{2a}{p}\right] + \left[\frac{3a}{p}\right] + \ldots + \left[\frac{p-1}{2}.\frac{a}{p}\right]\right\}.p$$

$$+ r_1 + r_2 + \ldots + r_n + s_1 + s_2 + \ldots + s_m$$

$$= t.p + r_1 + \ldots + r_n + s_1 + \ldots + s_m$$

Therefore, $$(a-1)\left(\frac{p^2-1}{8}\right) = a\left(\frac{p^2-1}{8}\right) - \frac{p^2-1}{8}$$

$$= \{t.p + (r_1 + \ldots + r_n) + (s_1 + \ldots + s_m)\}$$

$$- \{np - (r_1 + \ldots r_n) + (s_1 + \ldots. + s_m)\}$$

$$= p(t - n) + 2(r_1 + \ldots + r_n)$$

Therefore, since $p \equiv 1 \pmod 2$, we have

$$(a-1)\left(\frac{p^2-1}{8}\right) \equiv p(t-n) \equiv (t-n) \pmod 2$$

2.8. In this section, we prove that if p is an odd prime, then

$$\left(\frac{2}{p}\right) = (-1)^{\frac{p^2-1}{8}}$$

Proof. Take $a = 2$ in the previous section.

Here $$t = \left[\frac{2}{p}\right] + \left[\frac{4}{p}\right] + \left[\frac{6}{p}\right] + \ldots + \left[\frac{p-1}{2}.\frac{2}{p}\right]$$

$$= 0 + 0 + 0 + \ldots \ldots + 0$$

$$= 0$$

Since $(a-1)\left(\frac{p^2-1}{8}\right) \equiv (t-1) \pmod 2$, we have

$$\frac{p^2-1}{8} \equiv -n \pmod 2$$

Hence, $$\frac{p^2-1}{8} \equiv n \pmod 2.$$

But by Gauss's Lemma,

$$\left(\frac{2}{p}\right) = (-1)^n$$

Now, since $(-1)^{2k} = 1$, we have

$$(-1)^n = (-1)^{\frac{p^2-1}{8}}$$

and hence, $$\left(\frac{2}{p}\right) = (-1)^{\frac{p^2-1}{8}}$$

Example. $\left(\frac{2}{41}\right) = 1 = \left(\frac{2}{47}\right), \left(\frac{2}{227}\right) = -1 = \left(\frac{2}{229}\right)$

2.9. In this section we discuss the rules of Gauss in details regarding the mutual relations of quadratic residues. Let p and q be odd primes. Construct the rectangle ABCD by taking the points A = (0, 0), $B = \left(\frac{p}{2}, 0\right)$, $C = \left(\frac{p}{2}, \frac{q}{2}\right)$ and $D = \left(0, \frac{q}{2}\right)$ as vertices.

In this rectangle the points on the hypotenues AC are $\left(x, \frac{qx}{q}\right)$. In this case, p and q are unequal odd primes.

Therefore, for $0 < x < \frac{p}{2}$, if one of x and $\frac{qx}{p}$ is integer, the other one will not be an integer. A point within the rectangle is said to be an

Integral Lattice Point (ILP) if its base and perpendicular are both integers, Clearly, no point on AC is an ILP.

Let (x, y) be an ILP in the triangle ABC for which $x = r$ and $y = s$. For such a point, since $0 < r < \frac{qx}{p}$, we must have $0 < s < \frac{qr}{p}$. So, s can take any of the values 1, 2, 3, ..., $\left[\frac{qr}{p}\right]$. Therefore, those integral lattice points (x, y) inside the triangle ABC for which $x = t$ are $\left[\frac{qr}{p}\right]$ in number. But r can take the values 1, 2, 3, ..., $\frac{p-1}{2}$.

So, the number of integral lattice points inside the triangle ABC

$$= \left[\frac{q}{p}\right] + \left[\frac{2q}{p}\right] + \left[\frac{3q}{p}\right] + \ldots + \left[\frac{q-1}{2} \cdot \frac{q}{p}\right].$$

Similarly, the number of integral lattice points inside the triangle ADC

$$= \left[\frac{p}{p}\right] + \left[\frac{2p}{p}\right] + \ldots + \left[\frac{q-1}{2} \cdot \frac{p}{p}\right]$$

Again, if r takes any value from the set $\left\{1, 2, 3, \ldots, \frac{p-1}{2}\right\}$ and s takes any value from the set $\left\{1, 2, 3, \ldots, \frac{q-1}{2}\right\}$, then (r, s) will be an integral lattice point of ABCD. Thus, the number of integral points inside the rectangle ABCD

$$= \left\{\frac{p-1}{2}\right\} \times \left\{\frac{q-1}{2}\right\}$$

So, $\left\{\frac{p-1}{2}\right\} \times \left\{\frac{q-1}{2}\right\}$ = the number of ILP inside the rectangle ABCD = the number of ILP in the triangle ABC + the number of ILP in the triangle ADC

$$= \left\{\left[\frac{q}{p}\right] + \left[\frac{2q}{p}\right] + \ldots + \left[\frac{p-1}{2}\cdot\frac{q}{p}\right]\right\}$$
$$+ \left\{\left[\frac{p}{q}\right] + \left[\frac{2p}{q}\right] + \ldots + \left[\frac{(q-1)p}{2q}\right]\right\}.$$

For a, p, t and n described in section 2.6, we have proved that $(a-1)\dfrac{p^2-1}{8} \equiv (t - x) \pmod 2$. By taking $a = q$, since $q - 1 \equiv 0 \pmod 2$, we observe that $t \equiv n \pmod 2$. So, $\left(\dfrac{q}{p}\right) = (-1)^n = (-1)^t$.

Since $t = \left[\dfrac{q}{p}\right] + \left[\dfrac{2q}{p}\right] + \ldots + \left[\dfrac{p-1}{2}\cdot\dfrac{q}{p}\right]$, we get

$$\left(\frac{q}{p}\right) = (-1)^{\left[\frac{q}{p}\right] + \left[\frac{2q}{p}\right] + \ldots + \left[\frac{p-1}{2}\frac{q}{p}\right]}$$

and similarly, $$\left(\frac{p}{q}\right) = (-1)^{\left[\frac{p}{p}\right] + \left[\frac{2p}{p}\right] + \ldots + \left[\frac{q-1}{2}\frac{q}{p}\right]}$$

So, $$\left(\frac{p}{q}\right)\left(\frac{q}{p}\right) = (-1)^{\left[\frac{q}{p}\right] + \ldots \left[\frac{p-1}{2}\frac{q}{p}\right] + \left[\frac{p}{q}\right] + \ldots \left[\frac{q-1}{2}\frac{p}{p}\right]}$$
$$= (-1)^{\left(\frac{p-1}{2}\right)\left(\frac{q-1}{2}\right)}$$

Corollary: $$\left(\frac{q}{p}\right) = \begin{cases} -\left(\dfrac{p}{q}\right) \text{if } p \equiv 3 \equiv q \pmod 4, \\ \left(\dfrac{p}{q}\right) \text{if } p \equiv 1 \pmod 4, \text{ or, } q \equiv 1 \pmod 4. \end{cases}$$

i.e., if $p \equiv 3 \equiv q \pmod 4$ and if $x^2 \equiv p \pmod q$ has a solution, then $x^2 \equiv q \pmod p$ has no solution. On the other hand, if $p \equiv 1 \pmod 4$ or $q \equiv 1 \pmod 4$, then in order that $x^2 \equiv p \pmod q$ has a solution it is necessary and sufficient that $x^2 \equiv q \pmod p$ has a solution.

Rules for obtaining $\left(\frac{a}{p}\right)$: Let $q_j > 2$ be a prime number and

$$a = (-1)^e 2^{r_0} q_1^{r_1} q_2^{r_2} \dots q_s^{r_s}.$$

Then, write $\left(\frac{a}{p}\right)$ as

$$\left(\frac{-1}{p}\right)^e \left(\frac{2}{p}\right)^{r_0} \left(\frac{q_1}{p}\right)^{r_1} \left(\frac{q_2}{p}\right)^{r_2} \dots \left(\frac{q_j}{p}\right)^{r_j} \dots \left(\frac{q_s}{p}\right)^{r_s}$$

and in this put the values of $\left(\frac{-1}{p}\right), \left(\frac{2}{p}\right), \dots, \left(\frac{q_j}{p}\right)$

Again, since $\left(\frac{m}{p}\right)^2 = 1$, write 0 or 1 in places of e, r_0, ..., r_i, ..., according as they are even or odd respectively.

Example 1. $\left(\frac{-23}{41}\right) = \left(\frac{-1}{41}\right)\left(\frac{23}{41}\right)$

$$= \frac{23}{41} \qquad \left[\because \left(\frac{1}{41}\right) = (-1)^{\frac{41-1}{2}} = 1\right]$$

$$= \left(\frac{41}{23}\right) \qquad [\because 41 \equiv 1 \pmod 4)]$$

$$= \frac{18}{23} \qquad \left[\because \text{if } a \equiv b \pmod p\text{, then} \left(\frac{a}{p}\right) = \left(\frac{b}{p}\right)\right]$$

$$= \left(\frac{2}{23}\right)\left(\frac{3}{23}\right)^2 = \left(\frac{2}{23}\right) = (-1)^{\frac{23^2-1}{8}}$$

$$= (-1)^{\frac{24\times 22}{8}} = 1$$

i.e., it is possible to have solutions of $x^2 \equiv -23 \pmod{41}$.

Example 2. $\left(\frac{-59}{131}\right) = \left(\frac{-1}{131}\right)\left(\frac{59}{131}\right)$

$$= (-1)^{\frac{131-1}{2}} \left(\frac{59}{131}\right)$$

$$= (-1)\left(\frac{59}{131}\right) = (-1)\,(1)\,\frac{131}{59} \quad [\because\ 59 \equiv 3 = 131 \pmod 4]$$

$$= \left(\frac{13}{59}\right) \quad [\because\ 131 \equiv 3 \pmod{59}]$$

$$= \left(\frac{59}{13}\right) \quad [\because\ 13 \equiv 3 \pmod 4]$$

$$= \left(\frac{7}{13}\right) = \left(\frac{13}{7}\right) = \left(\frac{-1}{7}\right) = (-1)^{\frac{7-1}{2}} = -1.$$

So it is impossible to have solutions of $x^2 + 59 \equiv 0 \pmod{131}$.

Example 3. $\left(\frac{-189}{491}\right) = \left(\frac{-1}{491}\right)\left(\frac{3}{491}\right)^3\left(\frac{7}{491}\right)$

$$= (-1)\left(\frac{3}{491}\right)\left(\frac{7}{491}\right) = (-1)\left(\frac{491}{3}\right)\left(\frac{491}{7}\right)$$

$$[\because\ 491 \equiv 7 \equiv 3 \pmod 4]$$

$$= (-1)\left(\frac{2}{3}\right)\left(\frac{1}{7}\right) = (-1)\,.\,(-1)\,.\,(-1) = 1$$

So, it is possible to have solutions of $x^2 + 189 \equiv 0 \pmod{491}$.

Example 4. What are the odd primes p for which 3 is a quadratic residue.

Firstly, since $\left(\frac{3}{p}\right)\left(\frac{p}{3}\right) = (-1)^{\frac{3-1}{2}\cdot\frac{p-1}{2}}$

$$= (-1)^{\frac{1}{2}(p-1)},$$

$$\left(\frac{3}{p}\right) = \left(\frac{3}{p}\right)\left(\frac{p}{3}\right)^2 = (-1)^{\frac{1}{2}(p-1)} \cdot \left(\frac{p}{3}\right)$$

So, in order that $\left(\frac{3}{p}\right) = 1$, we must have

$$\left(\frac{p}{3}\right) = (-1)^{\frac{p-1}{2}}$$

But, $\left(\frac{p}{3}\right) \equiv p^{\frac{1}{2}(3-1)} = p \pmod{3}.$

So, in order that $\left(\frac{3}{p}\right) = 1$

we must have $p \equiv (-1)\ (p-1) \pmod{3}$

That is if, $p \equiv 1 \pmod{3}$ or, $p \equiv 4 \pmod{4}$, then it must be true that $p \equiv -1 \pmod{3}$. Again if $p \equiv 3 \pmod{4}$, then $p \equiv -1 \pmod{4}$. So,

in order that $\left(\frac{3}{p}\right) = 1$ we must have

$$p \equiv \pm 1 \pmod{12}$$

Example 5. What are the values of p for which (-3) is a quadratic residue.

Here, if $p > 3$, then

$$\left(\frac{-3}{p}\right) = \left(\frac{-1}{p}\right)\left(\frac{3}{p}\right) = (-1)^{\frac{p-1}{2}}\left(\frac{3}{p}\right)$$

$$= (-1)^{\frac{p-1}{2}} \left(\frac{3}{p}\right)\left(\frac{p}{3}\right)\left(\frac{p}{3}\right)$$

$$= (-1)^{\frac{p-1}{2}} . (-1)^{\frac{p-1}{2} \cdot \frac{3-1}{2}} \left(\frac{p}{3}\right)$$

$$= \left(\frac{p}{3}\right) = p^{1/2(3-1)} = p \text{ (mod 3), so}$$

$$\left(\frac{-3}{p}\right)\begin{cases} 1 \text{ if } p \equiv 1 \text{ (mod 3)} \\ -1 \text{ if } p \equiv -1 \text{ (mod 3)} \end{cases}$$

So, for $p = 2, 3$ and $p \equiv +1$ (mod 3), it is possible to have a solution of $x^2 \equiv -3$.

Example 6. What are the values of p for which (–2) is a quadratic residue?

Clearly, for this, we must have

$$1 = \left(\frac{-2}{p}\right) = \left(\frac{-1}{p}\right)\left(\frac{2}{p}\right) \quad \text{or,} \quad \left(\frac{-1}{p}\right) = \left(\frac{2}{p}\right) = \pm 1 \text{ or } -1$$

$$\text{But } \left(\frac{-1}{p}\right) = (-1)^{\frac{p-1}{2}} \quad \text{and} \quad \left(\frac{2}{p}\right) = (-1)^{\frac{p^2-1}{8}}$$

$$\text{So, } (-1)^{\frac{p-1}{2}} = (-1)^{\frac{p^2-1}{8}} \quad \text{or,} \quad (-1)^{\frac{p^2-1}{8} \cdot \frac{p-1}{2}} = 1$$

$$\text{or, } (-1)^{\frac{(p-3)(p-1)}{8}} = 1$$

So it is necessarily true that $p \equiv 3$ or 1 (mod 8) and this is sufficient also.

Example 7. If $p \neq 5$, then what is the value of $\left(\frac{5}{p}\right)$

Here, since

$$\left(\frac{5}{p}\right) = \left(\frac{5}{p}\right)\left(\frac{p}{5}\right)\left(\frac{p}{5}\right) = (-1)^{\frac{5-1}{2}}(-1)^{\frac{p-1}{2}}\left(\frac{p}{5}\right)$$

$$= \left(\frac{p}{5}\right)$$

$$= p^{1/2(5-1)} = p^2 \pmod 5,$$

$$\left(\frac{5}{p}\right) = \begin{cases} 1 \text{ if p } \equiv \pm 1 \pmod 5, \\ -1 \text{ if } p \equiv \pm\ 2 \pmod 5. \end{cases}$$

Example 8. The prime numbers which can be expressed in the form $(4n + 1)$ forms an infinite set.

Otherwise, assume that this set can be written as

$$p_1 < p_2 < p_3 < p_4 < \ldots < p_m$$

Then $\quad K = (2.\ p_1.\ p_2 \ldots p_m)^2 + 1 = 4n + 1$

will be divisible by an odd prime π,

i.e., $\quad (2p_1.\ p_2 \ldots p_m)^2 \equiv -1 \pmod \pi$

or, $\quad \left(\frac{-1}{\pi}\right) = 1$. Since

$$\left(\frac{-1}{\pi}\right) = (-1)^{1/2(\pi - 1)}, \text{ we must have}$$

$$\pi = 4k + 1.$$

But since $K \equiv -1 \pmod{p_j}$ and $K \equiv 0 \pmod \pi$, π is not equal to any p_j. So π is greater than p_m and is a prime number which can be expressed as $(4k + 1)$. This contradiction proves the result.

2.10. In this section we introduce the definition of *Jacobi symbol.* Let $(a, b) = 1$, b be an odd integer and for some prime numbers $p_1.\ p_2 \ldots p_r$, $b = p_1.\ p_2.\ p_3 \ldots p_r$. We define the Jacobi symbol $\left(\frac{a}{b}\right)$ as the product $\left(\frac{a}{p_1}\right)\left(\frac{a}{p_2}\right)\left(\frac{a}{p_3}\right)\ldots\left(\frac{a}{p_r}\right)$ of legendre symbols.

Since $(a, b) = 1$, a and b are unique in the definition of $\left(\frac{a}{b}\right)$.

If for two positive integers a', b', we have $(aa', bb') = 1$, then we leave it as an exercise for the reader to show that

(*i*) $\left(\frac{a}{bb'}\right)=\left(\frac{a}{b}\right)\left(\frac{a}{b'}\right)$

(*ii*) if $a = a'$ (mod b), then $\left(\frac{a}{b'}\right)=\left(\frac{a'}{b}\right)$

(*iii*) $\left(\frac{a'}{b}\right)=\left(\frac{a}{b}\right)\left(\frac{a'}{b}\right)$ (*iv*) $\left(\frac{a^2}{b}\right)=\left(\frac{a}{b^2}\right)=1$

(*v*) $\left(\frac{a^2a'}{b^2b'}\right)=\left(\frac{a'}{b'}\right)$ (*vi*) $\left(\frac{-1}{b}\right)=(-1)^{1/2(b-1)}$

(*vii*) $\left(\frac{2}{b}\right)=(-2)^{\frac{1}{8}(b^2-1)}$ (*viii*) $\left(\frac{a}{b}\right)\left(\frac{b}{a}\right)=(-1)^{\frac{a-1}{2}\cdot\frac{b-1}{2}}$

In order to prove (*vii*) and (*viii*) for odd integers

$u_1 . u_2 ... u_n$, we need to prove that

$$\frac{u_1-1}{2}+\frac{u_2-1}{2}+...+\frac{u_n-1}{2}\equiv\frac{u_1u_2...u_n-1}{2}(\text{mod}\,2)$$

and $$\frac{u_1^2-1}{8}+\frac{u_2^2}{8}+...+\frac{u_n^2-1}{8}\equiv\frac{(u_1u_2...u_n)^1-1}{8}(\text{mod } 2)$$

In order to prove these, it is enough to prove that

$$\frac{u+1}{2}+\frac{v-1}{2}\equiv\frac{uv-1}{2}\ (\text{mod } 2)$$

and $$\frac{u^2-1}{8}=\frac{v^2-1}{8}\equiv\frac{(uv)^2-1}{8}\ (\text{mod } 2)$$

It may be remarked here that if $x^2 \equiv a \pmod{b}$, then $\left(\frac{a}{b}\right)=1$. For if $x^2 \equiv a \pmod{b}$, then $x^2 \equiv a \pmod{p_j}$, $j = 1, 2, \ldots, n$ and $\left(\frac{a}{b}\right)=\left(\frac{a}{p_1}\right)\left(\frac{a}{p_2}\right)\cdots\left(\frac{a}{p_n}\right)$ = 1.1....1 = 1. But if $\left(\frac{a}{b}\right)=1$, then $x^2 \equiv a \pmod{b}$ may not have a solution such as: $\left(\frac{2}{33}\right)=\left(\frac{2}{3}\right)\left(\frac{2}{11}\right)$ = (–1) (–1) = 1, but $x^2 \equiv 2 \pmod{33}$ has no solution since $x^2 \equiv 2 \pmod{3}$ or $x^2 \equiv 2 \pmod{11}$ has no solution.

EXERCISE

1. Prove that each of 1, 4, 9, 16, 8; 2, 15 and 13 is a quadratic residue mod 17.
2. Prove that each of 1, 4, 9, 16, 17, 11, 7 and 5 is a quadratic residue mod 19.
3. Prove that each of 1, 4, 9, 16, 2, 13, 3, 18, 12, 8, and 6 is a quadratic residue mod 23.
4. Solve the following congruences:

   (*i*) $x^2 \equiv 1$ (*ii*) $x^2 \equiv 3$ (*iii*) $x^2 \equiv 4$ (*iv*) $x^2 \equiv 5$ (*v*) $x^2 \equiv 8$

   (*vi*) $x^2 \equiv 9$ (*vii*) $x^2 \equiv 10$ (*viii*) $x^2 \equiv 10$ (*ix*) $x^2 \equiv 12 \pmod{13}$

   Ans. (*i*) ±1, (*ii*) ±4, (*iii*) ±2, (*iv*) does not exist (*v*) does not exist, (*vi*) ±3, (*vii*) ±6, (*viii*) ±5.

5. Prove that each of $\left(\frac{-2}{19}\right), \left(\frac{-3}{19}\right), \left(\frac{6}{19}\right), \left(\frac{-1}{17}\right), \left(\frac{-2}{17}\right), \left(\frac{-3}{17}\right)$ and $\left(\frac{3}{17}\right)$ is equal to –1.

6. Prove that $\left(\frac{1}{p}\right)+\left(\frac{2}{p}\right)+\left(\frac{3}{p}\right)+\ldots+\left(\frac{p-1}{p}\right)=0.$

7. Show that $x^2 \equiv 2 \pmod{47}$ has two solutions, but $x^2 \equiv 2 \pmod{67}$ has no solution.

8. Determine the number of solutions of the following congruences:

(*i*) $x^2 \equiv 2 \pmod{94}$ (*ii*) $x^2 \equiv 2 \pmod{134}$

Ans. (*i*) 2, (*ii*) 0.

9. Prove that if for an odd prime p

$x^2 \equiv a \pmod{p}$ has a solution, then $x^2 \equiv a \pmod{p^r}$ has also a solution.

10. Find the values of the following Legendre symbols:

(*a*) $\left(\frac{2}{97}\right)$ (*b*) $\left(\frac{19}{23}\right)$ (*c*) $\left(\frac{-29}{67}\right)$

(*d*) $\left(\frac{-38}{29}\right)$ (*e*) $\left(\frac{-23}{59}\right)$ (*f*) $\left(\frac{-79}{97}\right)$

(*g*) $\left(\frac{94}{67}\right)$ (*h*) $\left(\frac{135}{229}\right)$ (*i*) $\left(\frac{-135}{229}\right)$

(*j*) $\left(\frac{2.3.5.7.11}{127}\right)$ (*k*) $\left(\frac{630}{53}\right)$ (*l*) $\left(\frac{-999}{73}\right)$

Ans. (*a*) 1, (*c*) –1, (*f*) 1, (*i*) 1.

In the following problems, p denotes an odd prime.

11. Prove that the product of quadratic residues mod p different from zero is congruent to $(-1)^{1/2(p+1)}$.

12. Determine the values of p for which $x^2 \equiv 11 \pmod{p}$.

Ans. $p = 1, 5, 7, 19, 25, 35, 37, 39, 43 \pmod{44}$.

13. Determine the values of p for which

(*a*) $\left(\frac{6}{p}\right) = 1,$ (*b*) $\left(\frac{10}{p}\right) = 1,$ (*c*) $\left(\frac{14}{p}\right) = 1$

Ans. (*b*) $p \equiv \pm 1, \pm 3, \pm 9, \pm 13 \pmod{40}$.

14. What are the values of p for which each of 2 and 3 is a quadratic residue?

Ans. $p \equiv \pm 1 \pmod{24}$.

15. Show that the set of all prime numbers which can be expressed in the form $(3n - 1)$ is an infinite set.

16. Show that $x^2 - 3 \equiv 0 \pmod{7^2.19^2.23}$ has no solution but $x^2 + 3 \equiv 0 \pmod{7^2.19^2.31}$ has exactly eight solutions.

17. Let $(a, b) = 1$, b an odd integer, and divisible by r distinct prime numbers and for each divisor p, $\left(\frac{a}{p}\right) = 1$.

Show that $x^2 \equiv a \pmod{b}$ has 2^r distinct solutions.

18. If $\left(\frac{a}{p}\right) = \left(\frac{b}{p}\right)$, then show that $ax^2 \equiv b \pmod{p}$ has a solution and if $\left(\frac{a}{p}\right) \not\equiv \left(\frac{b}{p}\right)$, then $ax^2 \equiv b \pmod{p}$ has no solution.

CHAPTER 3

General Properties of Congruence

3.1. Let $c_0, c_1, ..., c_n$ be integers and for some integer x, we have

$$f(x) = c_0x^n + c_1x^{n-1} + ... + c_n \equiv 0 \pmod{m}$$

Then, x is said to be a root of the congruence. If a is a root, then any integer which is congruent to a (mod m) is also a root and we identify all such roots with a.

The rules of algebra are not valid for finding out the roots of a congruence. For example, if p is a prime number, then

$$x^{p-1} \equiv 0 \pmod{p}$$

has $(p - 1)$ roots and these are 1, 2, 3, ..., $(p - 1)$.

But

$$x^4 - 1 \equiv 0 \pmod{16}$$

has eight roots such as: 1, 3, 5, 7, 9, 11, 13 and 15.

Again,

$$x^4 - 3 \equiv 0 \pmod{16}$$

does not have a single root.

3.2. Let $c_0, c_1, ..., c_n$ be integers and let x be an indeterminate.

Then, $c_0x^n + c_1x^{n-1} + ... + c_n$

is called an integral polynomial.

If

$$f(x) = c_0x^n + c_1x^{n-1} + ... + c_n,$$
$$g(x) = c_0'x^n + c_1'x^{n-1} + ... + c_n'$$

and for each $k = 0, 1, 2, ..., n$ we have $ck = ck' \pmod{m}$, then $f(x)$ is said to be congruent to $g(x)$ (mod m) and is written as:

$$f(x) \equiv g(x) \pmod{m}$$

It is clear that if $f(x) \equiv g(x)$, then for any integral polynomial $h(x)$, we have $$f(x)h(x) \equiv g(x)\ h(x).$$

If $f(x) \equiv g(x)h(x) \pmod{m}$, then we say that $g(x)$ divides $f(x) \pmod{m}$ and we write this as:

$$g(x) \mid f(x) \pmod{m}$$

Lemma:

$$(x - a) \mid f(x) \pmod{m},$$

if and only if

$$f(a) \equiv 0 \pmod{m}$$

Proof. If $(x - a) \mid f(x) \pmod{m}$, then there is an integral polynomial $h(x)$ such that

$$f(x) \equiv (x - a)\ h(x) \pmod{m}.$$

So, $$f(a) \equiv 0 \pmod{m}.$$

If $$f(a) \equiv 0 \pmod{m}, \text{ then}$$

$$f(x) \equiv f(x) - f\ (a) \pmod{m}.$$

But

$$f(x) - f(a) = c_0(x^n - a^n) + c_1(x^{n-1} - a^{n-1}) + \ldots + c_{n-1}(x - a)$$

$$= (x - a)\ \{\sum_{r=0}^{n-1} c_r(x^{n-r-1} + x^{n-r-2}a + \ldots + a^{n-r-1})\}$$

$$= (x - a)\ h(x)$$

So, $(x - a) \mid f(x) \pmod{m}$

3.3. Theorem 26. If p is a prime number and

$$f(x)g(x) \equiv 0 \pmod{p}, \text{ then}$$

$f(x) \equiv 0 \pmod{p}$ or $g(x) \equiv 0 \pmod{p}$

Proof. Assume that $f(x) \not\equiv 0 \pmod{p}$ and $g(x) \not\equiv 0 \pmod{p}$. Let $f_1(x)$ and $g_1(x)$ be the polynomials obtained respectively from $f(x)$

and $g(x)$ by deleting those terms whose coefficients are divisible by p. Let

$$f_1(x) = c_{1k}x^k + ..., c_{1k} \not\equiv 0 \pmod{p}$$

$$g_1(x) = c'_{1l}x + ..., c'_{1l} \not\equiv 0 \pmod{p}$$

Then,

$$f_1(x)g_1(x) = c_1kc'_1lx^{k+l}, + ...$$

and

$$clkc'_1l \not\equiv 0 \pmod{p}$$

Hence, $f(x)g(x) \not\equiv f_1(x)g_1(x) \not\equiv 0 \pmod{p}$

This contradiction completes the proof.

In a similar manner it can be proved that

Theorem 27. *If p is a prime number and for a positive integer α,*

$$f(x)g(x) \equiv 0 \pmod{p^{\alpha}}$$

and

$$f(x) \not\equiv 0 \pmod{p}, \text{ then}$$

$$g(x) \equiv 0 \pmod{p^{\alpha}}$$

Theorem 28. *For a prime number p if*

$$f(x) \equiv g(x)h(x) \pmod{p},$$

then every root of $f(x) \equiv 0$ (mod p) is either a root of $g(x) \equiv 0$ (mod p) or a root of $h(x) \equiv 0$ (mod p).

Proof. If a is a root of $f(x) \equiv 0 \pmod{p}$, then $f(a) = g(a)h(a) \equiv 0 \pmod{p}$. So, $p \mid g(a)h(a)$. Since p is a prime number, $p \mid g(a)$ or $p \mid h(a)$. Thus, a is a root of $g(x) \equiv 0 \pmod{p}$ or $h(x) \equiv 0 \pmod{p}$.

In the above theorem, it is essential that p must be a prime number. For if $p = 4$, then 4 is a root of $(x - 2)(x + 2) \equiv 0 \pmod{4}$, but this is neither a root of $x - 2 \equiv 0 \pmod{4}$ nor a root of $x + 2 \equiv 0 \pmod{4}$.

Theorem 29. *If the degree of $f(x)$ is n and if $f(x) \equiv 0$ (mod p) has more than n roots, then*

$$f(x) \equiv 0 \pmod{p}.$$

In other words, if

$$f(x) = c_0x^n + c_1x^{n-1} + \ldots + c_n,$$

then each c_k is divisible by p.

In order that this theorem will be useful p must be greater than n.

Proof. It follows from Theorem 6 that if

$$kx - l \equiv 0 \pmod{p}$$

has more than one solution, then $k \equiv 0 \equiv l \pmod{p}$.

Assume that the theorem is true for every polynomial of degree less than n. If $f(x)$ is of degree n and if $f(a) \equiv 0 \pmod{p}$, then according to the above lemma,

$$f(x) \equiv (x - a)g(x) \pmod{p}$$

and degree $g(x) \leq (n - 1)$. By Theorem 26, any root of $f(x) \equiv 0 \pmod{p}$ is a root of $(x - a) \equiv 0 \pmod{p}$ or of $g(x) \equiv 0 \pmod{p}$. If $f(x)$ has more than n roots, then $g(x)$ will have more than $(n - 1)$ roots. Now, since the theorem is true for $g(x)$, $g(x) \equiv 0 \pmod{p}$ is always true, *i.e.*, if

$$g(x) = d_0x^{n-1} + d_1x^{n-2} + \ldots + d_{n-1}, \text{ then}$$

p divides each of $d_0, d_1, \ldots, d_{n-1}$.

Since $\quad f(x) = (x - a)\, g(x)$, we have

$f(x) \equiv 0 \pmod{p}$, *i.e.*, p divides each c_k.

Theorem 30. *If deg $f(x) = n$ and $a_1, a_2, \ldots, a_n$ are the roots of $f(x)$, then*

$$f(x) = c_0(x - a_1)(x - a_2)\ldots(x - a_n) \pmod{p}.$$

The proof of this can be obtained from Theorem 29. We know that for a prime number p, the roots of

$$x^{p-1} \equiv 1 \pmod{p}$$

are 1, 2, 3, ... (p – 1). If d divides (p – 1), then we can prove

Theorem 31. $x^d \equiv 1$ *(mod p)*, $d \mid (p-1)$ *has d roots.*

Proof. If $(x^{p-1} - 1) = (x^d - 1).g(x)$ then

$$g(x) = x^{p-1-d} + x^{p-1-2d} + \ldots + x^d + 1$$

Here the number of roots of $g(x) \equiv 0$ does not exceed $(p - 1 - d)$. Again the number of roots of $x^{p-1} - 1 \equiv 0$ is $(p - 1)$. So, the number of roots of $x^d - 1 \equiv 0$ is not less than d. But this being not greater than d, is exactly d.

3.4. In previous sections, we have discussed only modolu prime numbers. In this section, we consider modolu any integer. First of all, we consider a linear congruence.

Let it be $$ax \equiv b \pmod{m}.$$

If $(a, m) = d$, then in order that the above congruence has a solution, d must divide b. It follows from Theorem 7 that the number of solutions of the above congruence is exactly d.

If ρ is a solution of $\frac{a}{d}x = \frac{b}{d}\left(\text{mod } \frac{m}{d}\right)$, then

$$\rho, \rho + \frac{m}{d}, \rho + 2\frac{m}{d}, \ldots, \rho + (d-1)\frac{m}{d}$$

are the solutions of the given congruence.

If $$\frac{a}{d} = a', \frac{b}{d} = b', \frac{m}{d} = m', \text{ then}$$

the given congruence can be written as

$$ax' = b' \text{ (mod } m')$$

Since a' and m' are coprime, we can choose two integers r and s such that

$$a'r + m's = 1$$

or $$a'r = 1 \text{ (mod } m')$$

So, by multiplying r to both sides of

$$a'x = b \text{ (mod } m')$$

and writing $b'r = c'$ the given congruence $ax \equiv b \pmod{m}$ can be written as:

$$x = c' \pmod{m'}$$

3.5. Let $a_1x \equiv b_1 \pmod{\mu_1}$, $a_2x \equiv b_2 \pmod{\mu_1}$, ..., $a_kx \equiv b_k \pmod{\mu_k}$ be a system of congruences and let $\mu_1, \mu_2, ..., \mu_k$ be coprime. In order that this system has a solution, (a_t, μ_t) must divide b_t for t = 1, 2, ..., k. If this condition is satisfied, we can write the system as follows:

$$x \equiv c_1 \pmod{m_1},\ x \equiv c_2 \pmod{m_2},\ ...\ x \equiv c_k \pmod{m_k}.$$

Here $m_1, m_2, ..., m_k$ are coprime, *i.e.*, if $m = m_1\ m_2\ ...\ m_k = m_1M_1 = m_2M_2 = ... = m_kM_k$, then for each t, m_t and M_t are coprime. So, we can choose n_i such that

$$n_tM_t \equiv 1 \pmod{m_i}$$

If
$$X = n_1M_1c_1 + n_2M_2c_2 + ... + n_kM_kc_k$$

then, for each t,

$$x \equiv n_tM_tc_t \equiv \pmod{m_t}$$

Therefore, for this x,

$$x \equiv c_1 \pmod{m_1},\ ...,\ x \equiv c_k \pmod{m_k}$$

If for some y,

$$y \equiv c_1 \pmod{m_1},\ ...,\ y \equiv c_k \pmod{m_k}, \text{ then}$$

for $t = 1, 2, ..., k$.

$$y \equiv c_t \equiv x \pmod{m_t}$$

i.e., $(y - x)$ is divisible by each of $m_1, m_2, ..., m_k$. Since $m_1, m_2, ..., m_k$ are coprime, $(y - x)$ must be divisible by m.

Thus,

$$x \equiv c_t \pmod{m_t}$$

is the one and only one solution.

3.6. Congruence of higher degree

Let $h_0, h_1, ..., h_n$ be integers and let

$$f(x) = h_0x^n + h_1x^{n-1} + ... + h_n$$

Let $m = m_1m_2\ ...\ m_k$, where $m_1, m_2, ..., m_k$ are pairwise coprime. We now attempt to find the solution of

$$f(x) \equiv 0 \pmod{m}$$

For this congruence, we have

$$f(x) \equiv 0 \pmod{m_t} \ (t = 1, 2, ..., k)$$

$$f(c_t) \equiv 0 \pmod{m_t} \ (t = 1, 2, ..., k)$$

We will determine a common solution x for

$$x \equiv c_1 \pmod{m_1},\ x \equiv c_2 \pmod{m_2},\ ...,\ x \equiv c_k \pmod{m_k}.$$

i.e., an x for which all the above congruences are simultaneously satisfied. This is a simultaneous congruence and by previous section, this has one and only one solution:

$$x \equiv n_1 M_1 c_1 + n_2 M_2 c_2 + ... + n_k M_k c_k$$

So, for $t = 1, 2, ..., k$,

$$f(x) \equiv f(c_t) \equiv 0 \pmod{m_t} \text{ holds,}$$

i.e., each of $m_1, m_2, ..., m_k$ divides $f(x)$.

Since $m_1, m_2, ..., m_k$ are coprime, their product $m_1 m_2 ... m_k$ will divide $f(x)$,

i.e., $f(x) \equiv 0 \pmod{m}$ holds.

Thus, we proved that if for $t = 1, 2, ..., k$, x is a common solution of

$$f(x) \equiv 0 \pmod{m_t}, \text{ then}$$

x is a solution of

$$f(x) \equiv 0 \pmod{m}.$$

Conversely, it is clear that each solution of

$$f(x) \equiv 0 \pmod{m}$$

is a common solution of

$$f(x) \equiv 0 \pmod{m_t} \ (t = 1, 2, ..., k).$$

This common solution can be obtained as follows:

First choose $c_1, c_2, ..., c_k$ such that

$$f(c_t) \equiv 0 \pmod{m_t},\ (t = 1, 2, ..., k).$$

Find out the unique (common) solution of the following simultaneous congruences:

$$x \equiv c_1 \pmod{m_1}$$
$$x \equiv c_2 \pmod{m_2}$$
$$\dots\dots\dots\dots\dots\dots\dots\dots$$
$$x \equiv c_k \pmod{m_k}$$

This is given by $\quad x \equiv n_1 M_1 c_1 + n_2 M_2 c_2 + \dots + n_k M_k c_k$

For this x, $f(x) \equiv 0 \pmod{m}$ holds.

Example 1. Solve the following simultaneous congruences:

$$x \equiv 1 \pmod{2},\ x \equiv 3 \pmod{5},$$
$$x \equiv 4 \pmod{7},\ x \equiv 6 \pmod{11},$$

In this case, $m = 2.3.7.11$

$$M_1 = 5.7.11,\ M_2 = 2.7.11,\ M_3 = 2.5.11,\ M_4 = 2.5.7$$
$$n_1 = 1,\ n_2 = -1,\ n_3 = 3,\ n_4 = 4,$$
$$n_1 M_1 \equiv 1 \pmod{2},\ n_2 M_2 \equiv 1 \pmod{5},$$
$$n_3 M_3 \equiv 1 \pmod{7},\ n_4 M_4 \equiv 1 \pmod{11},$$

So, the solution is

$$\begin{aligned} x &= n_1 M_1 c_1 + n_2 M_2 c_2 + n_3 M_3 c_3 + n_4 M_4 c_4 \\ &= 385 - 462 + 1320 + 1260 \\ &= 193 \pmod{770} \end{aligned}$$

Illustration: Let $c_1, c_2, \dots, c_k$ respectively have $N(c_1), N(c_2), \dots, N(c_k)$ values, then

$$n_1 M_1 c_1 + n_2 M_2 c_2 + \dots + n_k M_k c_k$$

will have $N(c_1)\ N(c_2) \dots N(c_k)$ values. In this case

$$f(x) \equiv 0 \pmod{m_t}$$

has $N(c_t)$ roots. So the number of roots of $f(x) \equiv 0 \pmod{m}$ is the product of the number of roots of $f(x) \equiv 0 \pmod{m_t}$.

Again if $p_1, p_2, \dots, p_k$ are prime numbers, and

$$m = p_1^{a_1}, p_2^{a_2}, ..., p_k^{a_k}, \text{ then}$$

$$m_t = p_t^{a_t}.$$

3.7. Let $m = p_1^{a_1}, p_2^{a_2}, ..., p_k^{a_k}$ and let $p_1, p_2, ..., p_k$ be prime numbers. Then, the solutions of

$f(x) \equiv 0 \pmod m$ completely depend on the solutions of

$$f(x) \equiv 0 \left(\text{mod } p_1^{a_1}\right), f(x) \equiv 0\left(\text{mod } p_2^{a_2}\right), ...,$$

$$f(x) \equiv 0\left(\text{mod } p_k^{a_k}\right).$$

Therefore, in this section, we discuss the solution of

$$f(x) \equiv 0 \pmod{p^a} \qquad ... \text{(A)}$$

for some prime number p. Each root of this congruence is congruent to some root of

$$f(x) \equiv 0 \pmod{m^{a-1}} \qquad ... \text{(B)}$$

Let β be a root of (B). We first determine the roots of (A) which are congruent to β mod p^{a-1}. Let x be such a root of (A), then

$$x = \beta \pmod{p^{a-1}}$$

i.e.,

$$x = p^{n-1}t + \beta$$

Now, $\quad f(\beta + p^{a-1}t) = f(\beta) + p^{a-1}t\, f'(\beta) + p^{2a-1}t^2\, \dfrac{f''(\beta)}{2!} + ...$

Since $a > 1$, $2a - 2 \geq a$, $3a - 3 \geq a$ and so on.

Again, if

$$f(x) = \Sigma\, h_\nu\, x^\nu, \text{ then}$$

$$f^{(l)}(x) = \Sigma\, h_\nu\, \nu(\nu - 1) ... (\nu - l + 1)x^{\nu-1}.\ \frac{\nu(\nu-1)...(\nu-l+1)}{l!},$$

being the coefficient of x^l in the expansion of $(1 + x)^\nu$, is an integer.

Again, since each h_ν is an integer, it follows that $\frac{f^{(l)}(\beta)}{l!}$ is an integer. So,

$$f(\beta + p^{a-1}t) \equiv f(\beta) = p^{a-1}t\ f(\beta) \pmod{p^a}$$

Hence, the value of t can be obtained from the congruence.

$$f(x) = f(\beta + p^{a-1}t) \equiv f(\beta) + p^{a-1}t\ f'(\beta) \equiv 0 \pmod{p^a}.$$

Case 1. Let $f'(\beta) \not\equiv 0 \pmod{p}$. Then, in order that $\beta + t\,p^{a-1}$ is a root of $f(x) \equiv 0 \pmod{p^a}$, it is necessary and sufficient that

$$f(\beta + t\,p^{a-1} f'(\beta) \equiv 0 \pmod{p^a})$$

i.e., p^a divides $t\,f'(\beta)\,p^{a-1} + f(\beta)$. So, it follows that p divides $t\,f'(\beta) + \frac{f(\beta)}{p^{a-1}}$.

Now since

$$f'(\beta) \not\equiv 0 \pmod{p}, \text{ we have}$$

$$t \equiv -\frac{f(\beta)}{p^{a-1}f'(\beta)} \pmod{p}$$

Hence, t has one and only one value. So, the number of roots of $f(x) \equiv 0 \pmod{p^a}$ is same as the number of roots of $f(x) \equiv 0 \pmod{p^{a-1}}$.

Case 2. If $f'(\beta) \equiv 0 \pmod{p}$, then

$$f(\beta + t\,p^{a-1}) \equiv f(\beta) \pmod{p^a}.$$

So if $f(\beta) \not\equiv 0 \pmod{p^a}$, then $f(\beta + t\,p^{a-1}) \equiv 0 \pmod{p^a}$ holds for no t. Hence, if $f(\beta) \not\equiv 0 \pmod{p^a}$, then $f(x) \equiv 0 \pmod{p^a}$ has no solution.

Case 3. If $f'(\beta) \equiv 0 \pmod{p}$ and $f(\beta) \equiv 0 \pmod{p^a}$, then for each t,

$$f(\beta + t\,p^{a-1}) \equiv f(\beta) \pmod{p^a}.$$

By taking $t = 1, 2, ..., p$ we obtain

p roots of $\qquad f(x) \equiv 0 \pmod{p^a}$

Example 2. Let p be a prime number and $f(x) = x^{p-1} - 1$. In this case each of 1, 2, 3, ..., $p - 1$ is a root of $f(x) = 0 \pmod{p}$. Let ξ be any one of these roots. Then

$$f'(\xi) \equiv (p-1)\,\xi^{p-2} \not\equiv 0 \pmod{p}$$

i.e., for given ξ, there is only one root of

$$f(x) \equiv 0 \pmod{p^2}$$

By taking $\xi = 1, 2, ..., p - 1$ successively, we observe that

$$f(x) \equiv 0 \pmod{p^2}$$

has exactly $(p - 1)$ roots.

By applying the above argument again and again, one can prove that

Theorem 32. *If p is a prime number, then for each a, the congruence*

$$x^{p-1} - 1 \equiv 0 \pmod{p^a}$$

has exactly (p – 1) roots.

CHAPTER 4

Algebraic Numbers

4.1. We have so far discussed only about integers. We now introduce rational numbers. Before defining this, let us take an informal approach to the problem. We know that the set of all integers, Z, is closed under addition, subtraction and multiplication of integers. But by dividing one integer by a non-zero integer we do not always get an integer. In fact Z with respect to usual addition and multiplication of integers is what is called a ring with unit element 1, moreover this is an integral domain with unit element 1, but is not what is called a field. We now seek for a field (the field Q of rational numbers) which contains an isomorphic image of Z as a subset. The following construction not only gives us rationals starting from integers, in general, gives a field starting with any integral domain which contains an isomorphic image of the given integral domain as a subset. We now proceed for the construction of the field Q.

Let M be the set of all ordered pairs (a, b), where $a, b \in$ Z and $b \neq 0$. One can write $\frac{a}{b}$ for (a, b) and call it a fraction. In M, define a relation $\sim$ as follows:

$(a, b) \sim (c, d)$ if and only if $ad = bc$.

This defines an equivalence relation on M.

1. If $(a, b) \in$ M, then $(a, b) \sim (a, b)$. Since $ab = ba$, hence $\sim$ is reflexive.

2. If (a, b), $(c, d) \in$ M and $(a, b) \sim (c, d)$, then $a\,d = bc$, hence $cb = da$ and so, $(c, d) \sim (a, b)$ which shows that $\sim$ is symmetric.

3. If (a, b), (c, d) $(e, f) \in$ M, $(a, b) \sim (c, d)$ and $(c, d) \sim (e, f)$, then $ad = bc$ and $cf = de$. Thus, $bcf = bde$ and since $bc = ad$, it follows that $adf = bde$ and hence, $af = be$. Thus, $(a, b) \sim (e, f)$ and the relation $\sim$ transitive.

Hence ~ divides M into mutually disjoint equivalence classes. Let $[a, b]$ be the equivalence class in M of (a, b) and let Q be the set of all such equivalence classes $[a, b]$, where $a, b \in Z$ and $b \neq 0$. This Q is the candidate for the field, we are seeking. In other words, a rational number is an equivalence class $[a, b]$ of (a, b), where a, $b \in Z$ and $b \neq 0$. One can also write $\frac{a}{b}$ for a rational number $[a, b]$ if there is no confusion. One should note that as ordered pairs (1, 2) and (2, 4) or as fractions $\frac{1}{2}$ and $\frac{2}{4}$ are distinct, but as rational numbers both are identical because both of them are in the same equivalence class. It remains to show that the set Q defined as above is a field. For this purpose, we define addition and multiplication in Q: we first dispose of the addition:

$$[a, b] + [c, d] = [ad + bc, bd]$$

Since $b \neq 0$ and $d \neq 0$, it follows that $bd \neq 0$ and hence $[ad + bc, \text{bd}] \in Q$. We now assert that this addition is well-defined, *i.e.*, if $[a, b] = [a', b']$ and $[c, d] = [c', d']$, then $[a, b] + [c, d] = [a', b'] + [c', d']$. Since $[a, b] = [a', b']$ and $[c, d] = [c', d']$, we have $ab' = ba'$ and $cd' = dc'$. We have to show that $[ad + bc, bd] = [a'd' + b'c', b'd']$.

Now,
$$\begin{aligned}(ad + bc)b'd' &= adb'd' + bcb'd' = ab'dd' + bb'cd' \\ &= ba'dd' + bb'dc' = bd(a'd' + b'c')\end{aligned}$$

and this is the desired equality. Clearly $[o, b]$ acts as a zero element for this addition, $[-a, b]$ as the additive inverse (negative) of $[a, b]$ and Q is a commutative group under this addition.

We now turn to the multiplication in Q. We define

$$[a, b]\ [c, d] = [ac, bd]$$

Since $b \neq 0$, $d \neq 0$ we have $bd \neq 0$ and hence $[ac, bd] \in Q$. In a manner very much similar to the case of addition, one can show that if $[a, b] = [a', b']$ and $[c, d] = [c', d']$, then $[a, b]\ [c, d] = [a', b']$, $[c', d']$. One can show that the non-zero elements of Q (*i.e.*, all the elements $[a, b]$ with $a \neq 0$) form a commutative group under multiplication where $[d, d']$ acts as the unit element and $[d, c]$ acts as the inverse of $[c, d]$ (note that since $c \neq 0$, $[d, c] \in Q$). Also, it is a routine verification that the distributive law holds in Q and having been established all these, we conclude that Q is a field.

All that remains to show that there is an isomorphism of Z into Q. First notice that for $x \neq 0$, $y \neq 0$ in Z, $[ax, x] = [ay, y]$ since $(ax)y = x(ay)$. Let us denote $[ax, x]$ by $[a, 1]$.

Define $\phi : Z \to Q$ by

$\phi(a) = [a, 1]$ for every $a \in Z$. Clearly, ϕ is 1-1 and

since,

$$\phi(a + b) = [a + b, 1] = [a, 1] + [b, 1] = \phi(a) + \phi(b)$$

$$\phi(ab) = [ab + 1] = [a, 1]\,[b, 1] = \phi(a)\,\phi(b),$$

ϕ is a homomorphism. Thus, ϕ is an isomorphism from Z into Q and this completes the proof.

Numbers are usually used to denote the lengths, weights, volumes, etc. If we take the length of any side of a square as a unit, then the length of the hypotenuse of that square cannot be expressed by a rational number. To prove this, let x denote the hypotenuse.

Then $$x^2 = 1^2 + 1^2 = 2$$

Suppose that $x = \frac{a}{b}$ where a and b are integers, coprime to each other and $b \neq 0$.

Then $$2 = x^2 = \frac{a^2}{b^2}$$

Hence, $$a^2 = 2b^2$$

The square of an odd integer is not divisible by 2. Thus, a is even, say $a = 2c$.

Hence, $$4c^2 = 2b^2 \quad \text{or} \quad b^2 = 2c^2$$

Thus, b is even, say $b = 2d$. Therefore, a and b cannot be coprime and this contradiction proves the assertion that $x = \frac{a}{b}$ is impossible.

But we can find infinitely many m and n such that

$$\frac{m^2}{n^2} < 2 < \frac{(m+1)^2}{n^2}, \text{ i.e.,}$$

x lies between $\dfrac{m}{n}$ and $\dfrac{m+1}{n}$.

Let $$\left(\frac{m}{n}\right)^2 < \left(\frac{m_1}{n_1}\right)^2 < \left(\frac{m_2}{n_2}\right)^2 < \left(\frac{m_3}{n_3}\right)^2 < \ldots < 2$$

and for each m_k and n_k,

$$\frac{m_k^{\;2}}{n_k^{\;2}} < 2 < \frac{(m_k+1)^2}{n_k^{\;2}}.$$

In this case, if k increases, the difference between $\dfrac{m_k}{n_k}$ and x decreases.

Thus, the definition of x can be determined by the increasing sequence $\left\{\dfrac{m_k}{n_k}\right\}$. Similarly, the definition of x can be determined by a decreasing sequence $\left\{\dfrac{m_k^{\;1}}{n_k^{\;1}}\right\}$.

Divide the set of all rational numbers into two classes L and R such that

(A) every element of L is less than every element of R.

(B) each rational number is either in L or in R, and LUR constitutes the whole set of rational numbers.

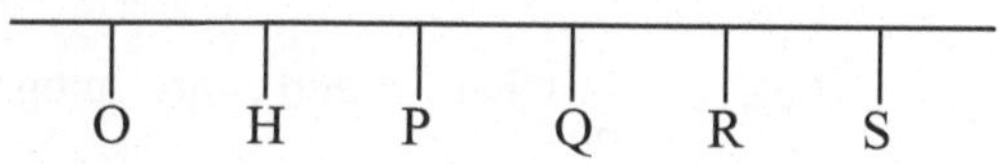

In the above figure, let O be a fixed point and let OP be a fixed length. If the lengths of OQ, OR, OS, ... are respectively the twice, thrice and four times of that of OP, then we represent the points P, Q, R, S, ... by 1, 2, 3, 4, ... respectively. Let us divide OS into 7 equal parts and if one part is equal to OH, then H represents the rational number $\dfrac{4}{7}$.

In this way all the positive rationals will be represented by some points on the line OS to the right of 0 and all the negative rationals will be represented be some points to the left of 0 on the extended line. But the length equal to $\sqrt{2}$ OP cannot be represented by any point in the above manner.

Firstly, let us represent all the rationals by points on the line OS. Let A be any point on that line. Those rationals which are represented by points to the left of A are members of the class L and those are represented by points to the right of A are members of R. If A itself is represented by a rational number, then it may belong to L or R. If A does not belong to L or R, then the number represented by the point A will be called an irrational number. For example, if $x^2 = 2$, then x is an irrational number. The number e is irrational and its proof is easy. π is also irrational, but its proof is somewhat difficult. For a simple and neat proof of this, one can see the paper by I. Niven entitled 'A simple proof that π is irrational' appeared in the Bulletin of the American Mathematical Society Vol. 53 (1947).

Any number which is either rational or irrational is called a real number. For more details of real numbers, the reader is referred to any book on Analysis. We also use complex numbers in this book and for details, the reader may consult any book on Complex Analysis. We prove only one theorem about real numbers here.

Theorem 33. *Let $c_1, c_2, ..., c_m$ be integers and x a real number such that*

$$x^m + c_1 x^{m-1} + c_2 x^{m-2} + ... + c_m = 0$$

Then x is either an integer or an irrational number.

Proof. Let $c_m \neq 0$. Let, $x = \frac{a}{b}$ where a and b are integers, coprime to each other and $b \neq 0$. Then,

$$a^m + c_1 a^{m-1}b + c_2 a^{m-2}b^2 + ... + c_m b^m = 0$$

So $b \mid a^m$. Let $a^m = kb$. If p is a prime factor of b, then since $p \mid b$ we have $p \mid a^m$. So $p \mid a$, *i.e.*, p is a common factor of a and b. Since a and b are coprime, this is impossible. In other words if $x = \frac{a}{b}$, then b cannot have a prime factor, *i.e.*, $b = 1$ and hence x is an integer.

4.2. Algebraic Numbers

Let c_0, c_1, ..., c_n be integers

and let $$c_0x^n + c_1x^{n-1} + ... + c_n = 0 \ (c_n \neq 0)$$

Suppose that there is a number ξ which satisfies the above equation, *i.e.*, when $x = \xi$,

$$c_0x^n + c_1x^{n-1} + ... + c_n = 0$$

Then, ξ is said to be a root of the above equation.

An algebraic number is a root of some equation having integer coefficients. If the equation is such that $c_0 = 1$, then a root of such an equation is called an *algebraic integer.*

A real number which is not algebraic is called a *transcendental number*. We now state a very important result known as *Gelfand-Schneider* Theorem. We cannot prove this theorem at this stage and will not try to prove the theorem in this introductory book.

Gelfand-Schneider Theorem. *Let* α *and* β *be algebraic numbers and let* β *be irrational. Then* α^{β} *is transcendental.*

While defining algebraic and transcendental numbers it was a partinent question to ask whether transcendental numbers exist. One and the best way of showing this is to demonstrate that some specific number is transcendental.

In 1851, *Liouville* gave a characterisation of a complex number to be algebraic. Using this, he was able to write down a large collection of transcendental numbers and this settled the question of existence. The next problem, which was of great importance, was whether some specific and familiar number was transcendental. The first success was due to *Hermite in* 1873 who proved that *e* is transcendental and his proof was simplified by Hilbert afterwards. The next success in this direction is probably due to *Lindemann* who in *1882* proved that π is transcendental. In 1934, working independently, Gelfand and Schneider proved the Gelfand-Schneider theorem which answered in the affirmative the question raised by Hilbert whether $2^{\sqrt{2}}$ was transcendental.

Examples of Algebraic Numbers

Example 1. Every rational number $\frac{a}{b}$ is an algebraic number, for if

$\xi = \frac{a}{b}$, then ξ is a root of the equation $bx - a = 0$.

Example 2. The complex number i is an algebraic number since this is a root of the equation $x^2 + 1 = 0$.

Example 3. There is a number w which is a root of the equation

$$x^2 + x + 1 = 0$$

Now, clearly

$$w^2 + w + 1 = 0$$

Again,

$$w^3 - 1 = (w - 1)(w^2 + w + 1) = 0$$

i.e., $w^3 = 1$ and $w^2 + w = -1$,

Hence, $x^2 + x + 1 = x^2 - (w + w^2)x + ww^2 = (x - w)(x - w^2)$

if $\xi = m + in$, then

$$(\xi - m)^2 = (i\ n)^2 = -n^2$$

or $\xi^2 - 2m\xi + m^2 + n^2 = 0$

Therefore, ξ is an algebraic integer. Such numbers are called **Gaussian integers.**

Again,

if $\eta = m + nw$, then

$$(\eta - m)^3 = (nw)^3 = n^3$$

So $\eta^3 - 3m\eta^2 + 5m\eta - m^3 + n^3 = 0$

Therefore, η is also an algebraic integer and these are called **Jacobian integers.**

It is now a pertinent question to ask whether there is any common feature between integers and the algebraic integers discussed above.

First of all, note that there is an essential difference between integers and rational numbers.

As remarked earlier Z with respect to usual addition and multiplication of integers is an integral domain with unit element 1, but Q with respect to usual addition and multiplication is a field.

Again, any integer can be expressed as the product of some prime numbers in a unique way, whereas such an expression is quite impossible for rationals. This difference is really an essential difference. Any integral domain with this property is known as an *Unique Factorisation Domain* (UFD). While claiming that the algebraic integers have some properties common with integers, we show that Gaussian integers and the Jacobian integers are Unique Factorisation Domains. Before proving this, we discuss some examples of integral domains and fields with respect to various types of addition and multiplication.

Example 1. As remarked earlier Z with usual addition and multiplication is an integral domain with unit element 1.

Example 2. Let $m > 1$ be an integer. Let D be the set of all integer multiples of m. D with usual addition and multiplication is an integral domain without unit element.

Example 3. Q and R with usual addition and multiplication are fields.

Example 4. Let Z_7 denote the set of integers mod 7 under the addition and multiplication mod 7. That is, $Z_7 = \{\overline{0}, \overline{1}, \overline{2}, \overline{3}, \overline{4}, \overline{5}, \overline{6}\}$, where $\overline{i} + \overline{j} = \overline{k}$, where k is the remainder of $i + j$ on division by 7, and $\overline{i} \,.\, \overline{j} = \overline{m}$, where m is the remainder of ij on division by 7. Thus, for instance,

$$\overline{4} + \overline{5} = \overline{2}, \overline{4}.\overline{6} = \overline{3} \text{ etc.}$$

In this case,

if

$$\overline{1}.\overline{1} = \overline{1} = \overline{6}.\overline{6}$$

$$\overline{2}.\overline{4} = \overline{1} = \overline{4}.\overline{2}$$

$$\overline{3}.\overline{5} = \overline{1} = \overline{5}.\overline{3}$$

Z_7 has seven elements and this is a finite integral domain.

Example 5. Let $m = 6$ and

$\overline{0}$ = [..., –30, –24, –18, –12, –6, 0, 6, 12, 18, 24, 30, ...]

$\overline{1}$ = [..., –29, –23, –17, –11, –5, 1, 7, 13, 19, 25, 31, ...]

$\overline{2}$ = [..., –28, –22, –16, –10, –4, 2, 8, 14, 20, 26, 32, ...]

$\overline{3}$ = [..., –25, –21, –15, –9, –3, 3, 9, 15, 21, 27, 33, ...]

$\overline{4}$ = [..., –24, –20, –14, –8, –2, 4, 10, 16, 22, 28, 34, ...]

$\overline{5}$ = [..., –23, –19, –13, –7, –1, 5, 11, 17, 23, 29, 35, ...]

In this case $\overline{n}.\overline{1} = \overline{n}$, but $\overline{2}.\overline{3} = \overline{0}$ and $\overline{2} \neq \overline{0}, \overline{3} \neq \overline{0}$. Thus, Z_6 is not an integral domain.

It is known from abstract algebra that *every finite integral domain is a field.* Using this, we prove that

If p is a prime number, then Z_p, the ring of integers mod p, is a field.

Proof. If $\overline{i}$, $\overline{j} \in Z_p$ and $\overline{i}\,.\,\overline{j} = \overline{0}$, then p must divide $i\,.\,j$. Since p is prime, it must divide i or j. Then, either $i \equiv 0 \pmod{p}$ or $j \equiv 0 \pmod{p}$, *i.e.*, either $\overline{i} = \overline{0}$ or $\overline{j} = \overline{0}$. Thus, Z_p is an integral domain.

Further, Z_p has p elements $\overline{0}, \overline{1}, \overline{2}, ..., \overline{(p-1)}$. So, by the above remark, Z_p is a field.

Example 6. Take $p = 11$. Then we get

The field, Z_{11} = $\{\overline{0}, \overline{1}, ..., \overline{10}\}$. In this case, $\overline{2}.\overline{5} = \overline{10} = \overline{3}.\overline{7}$.

Thus, we observe that in some integral domains (in some fields), one element cannot be expressed as the product of some elements in a unique way.

4.3. In this section, we investigate some integral domains which share the important property of integers that any positive integer can be expressed as the product of some integers in a unique way, in other words, which are Unique Factorisation Domains. Before investigating such integral domains, we introduce some definitions.

In an integral domain with unit element, an element which divides the unit element is called a *unit*.

For example, the elements 1 and –1 in Z divide the unit element 1 and hence these are the two units in Z. In any integral domain D, if e is a unit and ξ is any element of D, then ξ and $\xi.\ e$ are said to be *associates*. A non-unit in an integral domain is said to be a *prime* element if it is divisible by units or by its own associates and not divisible by any other element. In an integral domain D, a divides b means that there is an element $c \in$ D such that $a = bc$. In an integral domain, 'an element can be represented as a product of prime elements in a unique way' will mean that in any two different representations:

(*i*) the prime elements of one representation may be the associates of those of the second representation.

(*ii*) the order of prime elements or their associates may be different.

For example, the positive integer 24 is expressible uniquely as the product of prime numbers in the above sense although

$$\begin{aligned} 24 &= 2.2.2.3 = (-2).\ (-2).2.3 \\ &= 2.2.\ (-2).\ (-3) = (-3).2.2.(-2) \end{aligned}$$

If in an integral domain, we can identify prime elements and if any element can be represented uniquely as the product of prime elements in the above sense (*i.e.*, upto associates and order), then this is called a U.F.D.

We write

$$\text{Z} + i\text{Z} = \{m + in : m, n \in \text{Z}\}$$

$$\text{Z} + w\text{Z} = \{x + wy : x, y \in \text{Z}\}$$

We prove that Z + iZ and Z + wZ are U.F.D. and they are respectively called as Gaussian Domain and Jacobian Domain.

4.4. In this section we study some properties of Z + iZ. Let ξ, η and s with $s \neq 0$ be any three Gaussian integers. If $\xi = \eta s$, then ξ is said to be divisible by η, η is said to be a divisor of ξ and written as $\eta \mid \xi$.

Since 1, –1, i, –i are Gaussian integers, ξ has eight trivial divisors such as:

$$1, \xi, -1, -\xi, i, -i\xi, -i, i\xi$$

Let α, β, γ be Gaussian integers.

If $\alpha \mid \beta$ and $\beta \mid \gamma$, then $\alpha \mid \gamma$.

If $\alpha, r_1, r_2, ..., r_n, \beta_1, \beta_2, ..., \beta_n$ are Gaussian integers and if $\alpha \mid r_1$, $\alpha \mid r_2, ..., \alpha \mid r_n$, then

$$\alpha \mid \beta_1 r_1 + \beta_2 r_2 + ... + \beta_n r_n.$$

Any Gaussian integer which divides 1 is a unit of Z + iZ. For any Gaussian integer, $\xi = a + ib$, $(a^2 + b^2)$ is called the norm of ξ and is written as

$$a^2 + b^2 = N(\xi) = N(a + ib)$$

If $\overline{\xi} = a - ib$, the conjugate of ξ, then $N(\xi) = \xi\overline{\xi}$.

Further if $\eta = c + id \in Z + iZ$, then

$$N(\xi)\ N(\eta) = (a^2 + b^2)\ (c^2 + d^2)$$
$$= (ac - bd)^2 + (ad + bc)^2$$

and $\xi_\eta = (ac - bd) + i(ad + bc)$

Therefore,

$$N(\xi\eta) = N(\xi)\ N(\eta)$$

This result can be generalised to n Gaussian integers $\xi_1, \xi_2, ..., \xi_n$, *i.e.*,

$$N(\xi_1\ \xi_2 ... \xi_n) = N(\xi_1)\ N(\xi_2) ... N(\xi_n)$$

We prove

Theorem 34. *An element of Z + iZ is a unit if and only if its norm is 1.*

Proof. Let e be a unit. Then for some η,

$$1 = e\eta$$

So, $$1 = N(e\eta) = N(e)N(\eta)$$

Hence, N(e) divides 1. N(e) being the sum of squares of two integers N(e) = 1.

On the other hand, if N($a + ib$) = 1, then

$$1 = a^2 + b^2 = (a + ib)\,(a - ib)$$

So, $a + ib$ must divide 1 and hence $(a + ib)$ is a unit.

Theorem 35. *1, i, –1, i, are the only units of X + iZ.*

Proof. Let $a + ib$ be a unit.

Then, $$1 = \text{N}(a + ib) = a^2 + b^2$$

a and b being integers, $a^2 = 1$, or $b^2 = 1$.

So, $$a = \pm\, 1,\ b = 0 \text{ or } a = 0,\ b = \pm 1$$

So, $a + ib$ is one among 1, –1, i and $-i$.

If e is a unit and $\xi \in$ Z + iZ, then $e\xi$ is an associate of ξ.

Therefore, ξ, $-\xi$, $i\xi$ and $-i\xi$ are the only associates of ξ.

It is clear that if ξ divides η and if e_1 and e_2 are units, then e_1 divides e_2. In fact, any associate of ξ divides any associate of η.

Any nonzero element of Z + iZ which is not a unit is said to be a prime element if it is only divisible by its associates and by units. If π is a prime element of Z + iZ, then π is not divisible by any element except 1, π, –1, $-\pi$, i, $-i\pi$, $-i$, $i\pi$.

Theorem 36. *If the norm of a Gaussian integer is a prime number, then it must be a prime Gaussian number.*

Proof. Let p be a prime number and $\xi \in$ Z + iZ is such that N(ξ) = p. If $\xi = \eta s$ for η, $s \in$ Z + iZ, then

$$p = \text{N}(\xi) = \text{N}(\eta)\ \text{N}(s)$$

So either N(η) = 1 or N(s) = 1. Hence either η or s is a unit. So ξ is a prime Gaussian number.

Since N(2 + i) = 5, N(4 + i) = 17, N(5 + 2i) = 29, N(6 + 5i) = 61 and since 5, 17, 29, 61 as prime numbers, 2 + i, 4 + i, 5 + 2i, 6 + 5i are prime Gaussian numbers.

Remark. A prime number may or may not be a prime Gaussian number. Since

$$5 = (2 + i)(2 - i)$$

$$17 = (4 + i)(4 - i)$$

$$29 = (5 + 2i)(5 - 2i)$$

$$61 = (6 + 5i)(6 - 5i),$$

5, 17, 29 and 61 are not prime Gaussian numbers. But 7 is a prime Gaussian number.

Proof. Let a, b, c, d be integers such that

$$7 = (a + ib)(c - id).$$

Then, $7 = ac + bd$ and $0 = bc - ad$. Let $\frac{c}{a} = \frac{d}{b} = k$

Then, $7 = ac + bd = k(a^2 + b^2) = \frac{1}{k}(c^2 + d^2)$, that is, k is a positive number. If $k \leq 1$, then

$$7 = \frac{1}{k}(c^2 + d^2) \geq (c^2 + d^2)$$

But since $49 = N(7) = N(a + ib) \,.\, N(c + id) = (a^2 + b^2)(c^2 + d^2)$, $(c^2 + d^2)$ must divide 49. So $c^2 + d^2 = 7$ or 1.

But $$7 = 0 + 7 = 1 + 6 = 2 + 5 = 3 + 4.$$

Therefore, for $c \neq 0 \neq d$, $c^2 + d^2 \neq 7$. So $c^2 + d^2 = 1$. Similarly if $k \geq 1$, then $(a^2 + b^2) = 1$. So either $(c - id)$ or $(a + ib)$ is a unit. Thus, 7 is a prime element in Z + iZ. Using this fact, we can show that

Theorem 37. *Any prime number which is not the sum of two squares is prime, in Z + iZ. Clearly, any prime number p which can be expressed as the sum of two squares cannot be prime in Z + iZ because, if* $p = m^2 + n^2$, *then* $p = (m + in)(m - in)$ *and hence p cannot be prime in Z + iZ.*

Lemma. Any nonzero element of Z + iZ which is not a unit is divisible by some prime element of Z + iZ.

Proof. Let γ be a Gaussian integer and not a Gaussian prime number.

So, $$\gamma = \alpha_1\beta_1,\ N(\alpha_1) > 1,\ N(\beta_1) > 1$$

and $$N(\gamma) = N(\alpha_1)\ N(\beta_1)$$

So, $$1 < N(\alpha_1) < N(\gamma)$$

If α_1 is not a prime, then

$$\alpha_1 = \alpha_2\beta_2,\ N(\alpha_2) > 1,\ N(\beta_2) > 1$$

and $$N(\alpha_1) = N(\alpha_2)\ N(\beta_2)$$

So,

$$1 < N(\alpha_2) < N(\alpha_1)$$

Continuing in this way, we can get a Gaussian prime number α_s, such that

$$\gamma < \beta_1\alpha_1 = \beta_1\beta_2\alpha_2 = \beta_1\beta_2 \ldots \beta_s\alpha_s$$

and α_s will divide γ.

Theorem 38. *Any non-zero Gaussian integer which is not a unit can be expressed as a product of some Gaussian Prime numbers.*

Proof. Let γ be a nonzero Gaussian integer which is not a unit. This must be divisible by some Gaussian Prime number π_1, *i.e.,*

$$\gamma = \pi_1\gamma_1 \text{ and } N(\gamma_1) < N(\gamma)$$

If γ_1 is nonzero and not a unit, then for some Gaussian prime number π_2,

$$\gamma_1 = \pi_2\gamma_2 \text{ and } N(\gamma_2) < N(\gamma_1)$$

Continuing in this way, we get a decreasing set of positive integers, *i.e.,*

$$N(\gamma) > N(\gamma_1) > N(\gamma_2) > \ldots \geq 1.$$

Finally, there is some γ_k such that $N(\gamma_k) = 1$ and γ_k is a unit.

Therefore, $$\gamma = \pi_1\pi_2,\ldots\pi_{r-1}\gamma_k$$
$$= \pi_1\pi_2\ldots\pi_{r-2}\pi_{r-1}$$

and each of $\pi_1, \pi_2, \ldots, \pi_{r-2}$ and π_{r-1} is a Gaussian prime number.

Theorem 39. *If γ and $\gamma_1 \neq 0$ are two given Gaussian integers, then we can get Gaussian integers K and γ_2 such that*

$$\text{(A) } \gamma = K\gamma_1 + \gamma_2,$$

$$\text{(B) } N(\gamma_2) < N(\gamma_1).$$

Proof. Since $\gamma_1 \neq 0$,

$$\frac{\gamma}{\gamma_1} = \frac{\gamma\overline{\gamma_1}}{\gamma_1\overline{\gamma_1}} = R + iS \text{ (say)}$$

If $\gamma = a + ib$, $\gamma_1 = a_1 + ib_1$, then

$$R = (aa_1 + bb_1)/(a_1^2 + b_1^2).$$

$$S = (ab_1 - a_1b)/(a_1^2 + b_1^2)$$

R and S are rational numbers. if R is not an integer, let

$$m < R < m + 1$$

where m is an integer. Now one of $(R - m)$ and $(m + 1 - R)$ must be less than $\frac{1}{2}$. For if both are greater than $\frac{1}{2}$, we have

$$1 = (R - m) + (m + 1 - R) > \frac{1}{2} + \frac{1}{2} = 1$$

If $(R - m) < \frac{1}{2}$, let $m = x$, if $(m + 1 - R) < \frac{1}{2}$, let $m + 1 = x$, if $R - m = \frac{1}{2} = m + 1 - R$, let $m = x$. Thus, there is an integer x such that

$$-\frac{1}{2} \leq R - x \leq \frac{1}{2}$$

Similarly, there is an integer y such that

$$-\frac{1}{2} \leq S - y \leq \frac{1}{2}$$

$$N\left(\frac{\gamma}{\gamma_1} - x - iy\right) = (R - x)^2 + (S - y)^2 \leq \frac{1}{4} + \frac{1}{4} = \frac{1}{2}$$

Let $x + i.y = K$. If $\gamma_2 = \gamma - K\gamma_1$, then

$$N(\gamma_2)\ N(\gamma - K\gamma_1) = N\left(\frac{\gamma}{\gamma_1} - x - iy\right)N(\gamma_1)$$

$$\leq \frac{1}{2}N(\gamma_1) < N(\gamma_1)$$

We now adopt Euclid's algorithm.

Given γ, $\gamma_1 \neq 0$, we can get γ_2 such that for some Gaussian integer K,

$$\gamma = K\gamma_1 + \gamma_2,\ N(\gamma_2) < N(\gamma_1)$$

If $\gamma_2 \neq 0$, we can get Gaussian integers K_1 and γ_3 such that

$$\gamma_1 = K_1\gamma_2 + \gamma_3,\ N(\gamma_3) < N(\gamma_2)$$

Continuing in this way, we get Gaussian integers

$\gamma_4, \gamma_5, \ldots$ such that

$$N(\gamma) > N(\gamma_1) > N(\gamma_2) > \ldots \geq 0$$

Since $N(\gamma_1), N(\gamma_2), \ldots$ is a decreasing sequence, for some n, we must have

$$N(\gamma_{n+1}) = 0$$

So,

$$\gamma_{n+1} = 0,$$

$$\gamma_{n+1} = K_{n-1}\gamma_n + \gamma_{n+1}$$

$$\gamma_n = K_n\gamma_{n+1}$$

So γ_{n+1} will divide γ_n. So γ_{n+1} will divide $(K_{n-1}\gamma_n + \gamma_{n+1})$ $= \gamma_{n-1}$. Hence, γ_{n+1} will divide $(K_{n-2}\gamma_{n-1} + \gamma_n) = \gamma_{n-2}$ and consequently, γ_{n+1} will divide $(K_{n-3}\gamma_{n-2} + \gamma_{n-1}) = \gamma_{n-3}$.

Continuing in this way, we observe taht γ_{n+1} will divide γ and γ_1.

If δ divides γ and γ_1, then δ will divide $\gamma - k\gamma_1 = \gamma_2$. So, δ will divide $\gamma_1 - k_1\gamma_2 = \gamma_3$. Continuing in this way we, observe that δ divides γ and γ_1.

If s is a Gaussian integer such that every common factor of γ and γ_1 divides s and s divides γ and γ_1, then s is called the Highest Common Factor (HCF) of γ and γ_1 and is written as

$$s = (\gamma, \gamma_1)$$

If η and s are two Highest common factors, then

$$\eta \mid s \text{ and } s \mid \eta$$

So, $\quad s = \phi\eta,\ \eta = s = \theta\phi\eta$ or $\theta\phi = 1$

Theorem 40. *If* $(\gamma, \gamma_1) = 1$ *and if* $\gamma_1 \mid \beta\gamma$, *then* $\gamma_1 + \beta$.

Proof. It follows from Theorem 39 that

$$(\beta\gamma,\ \beta\gamma_1) = \beta\gamma_n$$

Since $(\gamma, \gamma_1) = 1$, γ_1 is a unit and

$$(\beta\gamma,\ \beta\gamma_1) = \beta$$

Now, $\gamma_1 \mid \beta\ \gamma_1$. Also, $\gamma_1 + \beta\gamma$. Therefore, from the definition of H.C.F., $\gamma_1 \mid (\beta\gamma, \beta\gamma_1)$. Thus, $\gamma_1 \mid \beta$.

Theorem 41. *If a Gaussian prime number* π *divides* $\beta\gamma$, *then* π *divides* β *or* γ.

Proof. Let $(\pi, \gamma) = \mu$. Then, μ divides π and γ. Hence since μ divides π,

(*i*) μ is a unit and $(\pi, \gamma) = 1$, or

(*ii*) μ is an associate of π and thus $(\pi, \gamma) = \pi$, that is, π divides γ.
In theorem 40, by taking $\gamma_1 = \pi$ we observe that

(*i*) if $(\pi, \gamma) = 1$, then $\pi \mid \beta$, or

(*ii*) $\pi \mid \gamma$.

Assume that a Gaussian integer γ can be expressed as product of Gaussian prime numbers in two different ways as follows:

$$\pi_1^{a_1}\ \pi_2^{a_2} \ldots \pi_k^{a_k}\ l = \gamma = p_1^{b_1}\ p_2^{b_2} \ldots p_e^{b_e}\eta$$

In this case, each π and p are primes, each a and b are positive integers and l and η are units.

π_i divides $p_1^{b_1}\ p_2^{b_2} \ldots p_e^{b_e}\ \eta$. If $ni \nmid p_1$, then π_i will divide

$$p_2^{b_2}\ p_3^{b_3} \ldots p_e^{b_e}\ \eta.$$

If π_i does not divide any p_j, then $\pi_i \mid \eta$ and hence π_i will not be prime. Since π_i is prime, it must divide some p_j. Here π_i and p_i are associates. If $b_i > a_i$, then p_j will divide

$$\pi_1^{a_1} \ldots \pi_{i-1}^{a_{i-1}} \; \pi_{i+1}^{a_{i+1}} \ldots \pi_k^{a_k}$$

$$= p_1^{b_1} \; p_i^{b_j - a_j} \ldots p_e^{b_e}$$

But each of $\pi_1,\ldots,\pi_{i-1}, \pi_{i+1},\ldots, \pi_k$ being coprime with p_j, this is impossible. So $b_j \not> a_j$. Similarly, it can be shown that $a_j \not> b_j$. Thus we observe that

(*i*) Each π is an associate of some p and so $k = e$,

(*ii*) if π and p are arranged in such a way that for $j = 1, 2, \ldots, k$, p_j is the associate of π_j, then $a_j = b_j$. Thus, we have proved that

Theorem 42. *Z + iZ is a Unique Factorisation Domain.*

4.5. Properties of Jacobian Domain Z + *w*Z. The units, associates and prime elements of Z + *w*Z can be defined in the same manner as in Z + *i*Z.

But if m and n are integers, the norm of $\xi = m + wn$ is given by

$$\begin{aligned} N(\xi) &= N(m + wn) = (m + wn)\,(m + w^2 n) \\ &= m^2 - mn + n^2 \end{aligned}$$

Since $m^2 - mn + n^2 = (m - \tfrac{1}{2}n)^2 + \tfrac{3}{4}n^2$, if $\xi \neq 0$, then $N(\xi)$ is a positive integer. If $\alpha = m_1 + wn_1$ and $\beta = m_2 + wn_2$, then since $1 + w + w^2 = 0$,

$$\begin{aligned} N(\alpha).\ N(\beta) &= [(m_1 - \tfrac{1}{2}m_1)^2 + \tfrac{3}{4}n_1^{\,2}]\ [(m_2 - \tfrac{1}{2}n_2)^2 + \tfrac{3}{4}n_2^{\,2}] \\ &= \{(m_1 - \frac{1}{2}n_1) + i\sqrt{\frac{3}{4}}n_1\}\ \{(m_1 - \frac{1}{2}n_1) - i\sqrt{\frac{3}{4}}n_1\} \text{X} \\ &\quad \{(m_2 - \frac{1}{2}n_2) + i\sqrt{\frac{3}{4}}n_2\}\ \{(m_2 - \frac{1}{2}n_2) - i\sqrt{\frac{3}{4}}n_2\} \\ &= [(m_1 m_2 - n_1 n_2) - \tfrac{1}{2}(n_1 m_2 + n_2 m_1) + \tfrac{1}{2}n_1 n_2]^2 + \\ &\qquad \tfrac{3}{4}[(n_1 m_2 + n_2 m_1) - n_1 n_2]^2 \\ &= [(m_1 m_2 - n_1 n_2) - \tfrac{1}{2}(n_2 m_1 + n_1 m_2 - n_1 n_2)]^2 \\ &\qquad + \tfrac{3}{4}(m_1 n_2 + m_2 n_1 - n_1 n_2)]^2 \end{aligned}$$

So $\quad N(\alpha\beta) = N(\alpha)N(\beta)$

As before we can prove

Theorem 43.

(*i*) *A Jacobian number is a unit if and only if its norm is 1,*

(*ii*) *A Jacobian number whose norm is 1 is a Jacobian prime number,*

(*iii*) *every non-zero Jacobian number which is not a unit is divisible by a Jacobian prime number,*

(*iv*) *every non-zero Jacobian number which is not a unit can be expressed as a product of Jacobian prime numbers.*

For a Jacobian unit $\quad a + wb,\ a^2 - ab + b^2 = 1$

or, $\quad (2a - b)^2 + 3b^2 = 4$

So, $(2a - b)^2 = 4 \quad$ or $\quad (2a - b)^2 = 1$

$3b^2 = 0 \qquad 3b^2 = 3$

That is, $a = \pm 1$ and $b = 0$, or, $a = 0$ and $b = \pm 1$

or $\quad a = 1$ and $b = 1$, or, $a = -1$ and $b = -1$

So, the units of Z + wZ are $\pm 1, \pm w, \pm(1 + w)$, *i.e.*, $\pm 1, \pm w, \pm w^2$. Again,

Theorem 44. *If* γ *and* $\gamma \neq 0$ *are two Jacobian numbers, we can choose two Jacobian numbers k and* γ_2 *such that*

(*i*) $\gamma = k\gamma_1 + \gamma_2$ and $\quad$ (*ii*) $N(\gamma_2) + N(\gamma_1)$

Proof. For integers a, b, c and d, we have

$$\frac{\gamma}{\gamma_1} = \frac{a + wb}{c + wd} = \frac{(a + wb)(c + w^2 d)}{(c + wd)(c + w^2 d)}$$

$$= \frac{ac + bd - ad}{c^2 - cd + d^2} + \frac{bc - ad}{c^2 - cd + d^2} w$$

$$= R + Sw \text{ (say)}$$

We can get two integers R and S such that

$$-\frac{1}{2} \le R - x \le \frac{1}{2}, -\frac{1}{2} \le s - y \le \frac{1}{2}$$

So, as before, by taking $k = x + wy$ we get

$$N\left(\frac{\gamma}{\gamma_1} - (x + wy)\right) = (R - x)^2 - (R - x)(S - y) + (S - y)^2 \leq \tfrac{3}{4} < 1$$

Now, if we put $\gamma_2 = \gamma - k\gamma_1$, we get

$$N(\gamma_2) = N(\gamma - k\,\gamma_1)) \leq \tfrac{3}{4}\,N(\gamma_1) < N(\gamma_1)$$

Now, as in the case of Z + *i*Z we can prove

Theorem 45. *Z* + wZ *is a U.F.D.*

We now prove that $\lambda - 1 = w$ is a prime element of Z + *w*Z. We have

$$N(\lambda) = N(1 - w) = 3$$

Since 3 is a prime number, λ is a prime Jacobian number.

If $\gamma = a + wb$, then $\gamma = a + b - b\lambda$,

If we divide $(a + b)$ by $3 = (1 - w)(1 - w^2) = \lambda(1 - w^2)$, then we get 0, 1 or 2 as the remainder.

So, $$\gamma = (a + b) \pmod{\lambda} = 0,\ 1 \text{ or } 2 \pmod{\lambda}$$

Again, since N(1) = 1, N(2) = 4 and N(λ) = 3, N(λ) does not divide N(1) or N(2). So, 0, 1 and 2 are not congruent to each other mod λ. Again,

$$\lambda^2 = 1 - 2w + w^2 = 3w$$

So, 3 and λ^2 are associates. Similarly, since

$$\lambda w^2 = -1 + w^2 = -1(1 - w^2)$$

$$\lambda w = w(1 - w)$$

$$\lambda = 1 - w$$

$\pm(1 - w)$, $\pm w(1 - w^2)$, $\pm(1 - w)$ are associates of λ. So

Theorem 46.

(*i*) $\lambda = 1 - w$ *is prime,*

(*ii*) $m + wn = 0, 1, -1 \pmod{\lambda}$,

(*iii*) *3 and* λ^2 *are associates,*

(*iv*) $\pm(1 - w)$, $\pm(1 - w^2)$, $\pm w(1 - w)$ *and* λ *are associates to each other.*

CHAPTER 5

Quadratic Fields

Integral Domain and Properties of their Primes.

5.1. In the previous chapter, we have discussed about the algebraic numbers. If $a_0, a_1, \ldots\ldots, a_n$ are integers and $a_0 \neq 0$, then each root of

$$a_0 x^n + a_1 x^{n-1} + \ldots\ldots + a_n = 0 \qquad \ldots(5.1)$$

is an algebraic number.

Let θ be an algebraic number and let $\{p_0, p_1, p_2, \ldots\ldots, p_k\}$ and $\{q_0, q_1, q_2, \ldots\ldots, q_l\}$ be two sets of rational numbers. We now construct two polynomials:

$$p(\theta) = p_0\theta^k + p_1\theta^{k-1} + \ldots + p_{k-1}\theta + p_k$$

and

$$Q(\theta) = q_0\theta^l + q_1\theta^{l-1} + \ldots\ q_{l-1}\theta + q_l$$

If $Q(\theta) \neq 0$, then let $\quad R(\theta) = \dfrac{P(\theta)}{Q(\theta)}$

Let $K(\theta)$ denote the set of all $R(\theta)$. This is called an *algebraic field.*

Let θ be a root of the nth degree equation (5.1) and is not a root of any equation whose degree is less than n. (Here, the coefficients of every equation are assumed to be integers). Then, θ is said to be an algebraic number of degree n.

If $n = 1$, then θ is a rational number and $K(\theta)$ is the field of rational numbers. We denote this field by $K(1)$.

If $n = 2$, then θ is said to be a quadratic number. In this case, θ is a root of a quadratic equation $a_0x^2 + a_1x + a_2 = 0$.

Therefore, $\quad \theta = \dfrac{-a_1 + \sqrt{a_1^2 - 4a_0a_2}}{2a_0}$

In order that $\sqrt{a_1^2 - 4a_0a_2}$ is an integer, $(a_1^2 - 4a_0a_2)$ must be the square of some integer. If $(a_1^2 - 4a_0a_2)$ contains squares of some integers as factors, then we can bring them out of the radical sign and hence θ can be expressed as:

$$\theta = \frac{a + b\sqrt{m}}{c}$$

Where m does not contain square of any integer as a factor. Such an m is called square free.

Since $$\sqrt{m} = \frac{c\theta - a}{b},$$

the two fields K(θ) and $K(\sqrt{m})$ are identical. In this chapter, we discuss the properties of $K(\sqrt{m})$. We assume that $m \neq 1$. If ξ is any element of $K(\sqrt{m})$, then ξ can be written as:

$$\xi = \frac{P(\sqrt{m})}{Q(\sqrt{m})} = \frac{t + u\sqrt{m}}{v + w\sqrt{m}}$$

$$= \frac{t + u\sqrt{m}}{v + w\sqrt{m}} \frac{v - w\sqrt{m}}{v - w\sqrt{m}} = \frac{a + b\sqrt{m}}{c}$$

In this case t, u, v, w. a. b *and* c are integers.

Thus, $$(c\xi - a)^2 = mb^2$$

i.e., ξ is a root of

$$c^2x^2 - 2acx + a^2 - mb^2 = 0$$

Hence, if $a^2 - mb^2 \neq 0$, then ξ is a quadratic number and if $a^2 - mb^2 = 0$, then ξ is a rational number.

5.2. Given any algebraic number ξ, it is natural to ask: what are the equations of which ξ is a root.

Firstly, let us consider the case of a rational number ξ, for example, take ξ = ½.

This can be a root of several equations given below:

(A) $2x - 1 = 0$

$4x - 2 = 0$

$6x - 3 = 0$

$\vdots$

(B) $2x^2 + x - 1 = 0$

$2x^3 + x^2 + x - 1 = 0$

... ...

$2x^4 + x^3 + x^2 + x - 1 = 0$

$2x^5 + x^4 + x^3 + x^2 + x - 1 = 0$, etc.

Of course, the equations given in (A) are not different from each other. For example, by dividing by 2 of both sides of the equation $4x - 2 = 0$ we get $2x - 1 = 0$. Similarly, every equation given in (A) is not different from $2x - 1 = 0$. So, in order to find the equation for some algebraic number, it is essential that the coefficients must be divided by the common factor, so that we do not get different forms of the same equation. In future we shall assume that the G.C.F. of the coefficients of every equation is 1.

Theorem 47. *Let ξ be an element of degree n and let $f(x)$ be a polynomial of degree n, whose coefficients do not have any common factor, satisfying $f(\xi) = 0$. If ξ is a root of any equation $g(x) = 0$, then for some polynomial $h(x)$, we must have*

$$g(x) = f(x)\ h(x).$$

Proof. First of all, it is clear that the degree of $g(x)$ is not less than that of $f(x)$. Therefore, we can write

$$g(x) = f(x).\ q(x) + r(x)$$

where $\deg r(x) < \deg f(x)$. By putting $x = \xi$, we obtain

$$0 = 0\ .\ q(\xi) + r(\xi) = r(\xi)$$

Thus, ξ is a root of $r(x) = 0$. But, since $\deg r(x) < \deg f(x)$, this is impossible. Therefore, $r(x) = 0$ and consequently,

$$g(x) = f(x).\ q(x)$$

Note that the coefficients of $g(x)$ and $f(x)$ are integers, but those of $q(x)$ may be any rational number. Let c be the L.C.M. of the denominators of the coefficients of $q(x)$. Now, we have

$$cg(x) = f(x).\ cq(x) = f(x).\ Q(x) \quad \text{(say)}$$

and in this case the coefficients of $Q(x)$ are rational numbers. Let the G.C.F. of the coefficients of $Q(x)$ be d. Then $d \mid c$. By dividing both sides of the above equation by d, we obtain

$$eg(x) = f(x).\ q_1(x)$$

Let e be divisible by a prime number p. Then,

$$f(x)\ .\ q_1(x) = 0\ (mod\ \text{p})$$

Therefore, by theorem 26, either $f(x) = 0 \pmod{p}$ or $q_1(x) = 0 \pmod{p}$. But this is impossible since the G.C.F. of the coefficients of $f(x)$ is 1 and G.C.F. of coefficients of $q_1(x)$ is 1. So, e is not divisible by any prime number p. Hence, $e = 1$ and by writing $h(x)$ for $q_1(x)$ we get

$$g(x) = f(x)\ h(x)$$

and this completes the proof.

Again if ξ is an element of degree n, $f(x)$ and $g(x)$ are two polynomials each of degree n and having no common factors among the coefficients and satisfying $f(\xi) = 0$, $g(\xi) = 0$, then $f(x)$ and $g(x)$ are not different. To prove this note that by Theorem 47, we have

$$g(x) = f(x)\ .\ h(x)$$

Since deg $g(x)$ = deg $f(x)$, the degree of $h(x)$ must be zero, *i.e.*, h(x) must be an integer c_1.

So, $$g(x) = c_1 f(x)$$

Similarly, for some integer c_2, we must have

$$f(x) = c_2 g(x)$$

Therefore, $$g(x) = c_1 f(x) = c_1 c_2 g(x)$$

and so, $c_1 c_2 = 1$. Since c_1 and c_2 are both integers,

$$c_1 = 1 = c_2 \qquad \text{or} \qquad c_1 = -1 = c_2$$

So, $f(x) = g(x)$ or $f(x) = -g(x)$

If in both $f(x)$ and $g(x)$, the coefficient of x_n is taken to be positive, then $f(x)$ and $g(x)$ will not be different.

An *algebraic integer* is a root of an equation with integer coefficients having 1 as the leading coefficient.

Theorem 48. *Each algebraic number of degree n is a root of a unique polynomial f(x) of degree n having no common factor among the coefficients and the coefficient of x^n being positive. If ξ is an algebraic integer, then the coefficient of x_n in f(x) is 1.*

If ξ is an algebraic number, then ξ is a root of some polynomial

$$x^j + c_1x^{j-1} + c_2x^{j-2} + \ldots + c_j = 0$$

If the left side of the above equation is denoted by g(x), then by Theorem 47,

$$g(x) = f(x)\ h(x)$$

In this case, if $f(x) = a_0x^n + \ldots$ and $h(x) = h_0x^{j-n} + \ldots$, *then* $a_0h = 1$. *So,* $a_0 = 1$ *and this proves the second part of Theorem 48.*

5.3. Suppose that the square of any integer is not a factor of m and $m \neq 1$. The algebraic integers which are in $K(\sqrt{m})$ are called the algebraic integers of $K(\sqrt{m})$. Each such algebraic integer is a root of some equation

$$x^2 + px + q \text{ } (p, q \text{ integers}).$$

Let a, b and c be integers, $c > 0$ and let the G.C.F. of a, b and c be 1. Let

$$\xi = \frac{a + b(\sqrt{m})}{c}$$

be an algebraic integer. If $b = 0$, then $\xi = \frac{a}{c}$ (a rational number) so $c = 1$ and $\xi = a$ (an integer). If $b \neq 0$, then ξ will be quadratic.

We know that in the equation

$$c^2x^2 - 2acx + a^2 - mb^2 = 0$$

For ξ, $a^2 - mb^2 \neq 0$; if $a^2 - mb^2 = 0$, then $b^2 \mid a^2$

and hence $m = \dfrac{a^2}{b^2}$ is a square.

Now, the algebraic integer ξ is a root of

$$x^2 - \frac{2a}{c}x + \frac{a^2 - mb^2}{c^2} = 0.$$

So, $c \mid 2a$ and $c^2 \mid (a^2 - mb^2)$. If $(a, c) = d$, then d^2 divides each of a^2, c^2, so $(a^2 - mb^2)$ and hence d^2 divides $mb^2 = a^2 - (a^2 - mb^2)$. Since m does not contain square of any integer as a factor, d will divide b. So d will divide a, b and c. But we have assumed that the G.C.F. of a, b and c is 1, so $d = 1$,

i.e., a and c are coprime. But $c \mid 2a$, so $c \mid 2$ and hence $c = 1$ or 2.

If $c = 2$, then since $(a, c) = 1$, a is an odd integer and since $4 = c^2$ divides $(a^2 - mb^2)$,

$$mb^2 \equiv a^2 \equiv 1 \pmod 4$$

So, m and b are both odd. So

$$b^2 \equiv 1 \pmod 4 \text{ and } m \equiv 1 \pmod 4$$

(1) If $m \not\equiv 1 \pmod 4$, then $c = 1$ and the elements of $K(\sqrt{m})$ are $a + b\sqrt{m}$. In this case, $m \equiv 2$ or $m \equiv 3 \pmod 4$ since m is not divisible by $4 = 2^2$.

(2) If $m \equiv 1 \pmod 4$, then take

$1 = \frac{1}{2}(\sqrt{m-1})$. By taking $m = 4n + 1$, we get

$(2p + 1)^2 = m = 4n + 1$ and hence, $p^2 + p = n$

Thus, p is an algebraic integer of $K(\sqrt{m})$.

If $c = 2$, since a and b are both odd,

$$\xi = \frac{a + b\sqrt{m}}{2} = \frac{a + b}{2} + bp$$

By taking $\frac{a+b}{2} = a_1$ and $b = 2b_1 + 1$, we observe that ξ can be expressed as:

$$\xi = a_1 + (2b_1 + 1)p$$

In this case, a_1 and b_1 are integers.

If $c = 1$, then

$$\xi = a + b\sqrt{m} = a + b + 2bp = a_1 + 2b_1 p, \text{ Hence}$$

Theorem 49. *The algebraic integers of $K(\sqrt{m})$ can be expressed as $a + b\sqrt{m}$ or $a + bp$, where a and b are integers.*

(A) *If $m \equiv 2$ or $m \equiv 3$ (mod 4), then any algebraic integer of $K(\sqrt{m})$ can be written as $a + b\sqrt{m}$.*

(B) *If $m \equiv 1$ (mod 4), then any algebraic integer of $K(\sqrt{m})$ can be written as $a + bp$.*

Again, let a and b be any two integers. Then, for $m \equiv 2$ or 3(mod 4), $a + b\sqrt{m}$ is an algebraic integer of $K(\sqrt{m})$ and for $m \equiv 1$ (mod 4), $a + bp$ is an algebraic integer of $K(\sqrt{m})$.

In other words, if we let $I(\sqrt{m})$ denote the set of all algebraic integers of $K(\sqrt{m})$, then

$$I(\sqrt{m}) = \begin{cases} Z + \sqrt{m}\,Z & \text{if } m \equiv 2 \text{ or } 3 \\ Z + pZ & \text{if } m \equiv 1 \end{cases}$$

5.4. We now define units, prime elements for the algebraic integers of $K(\sqrt{m})$.

If α and β are algebraic integers such that for some algebraic integer γ, $\alpha = \beta\gamma$, then β is said to divide α. An algebraic integer which divides 1 is called a unit (unity) of $I(\sqrt{m})$. If e is a unit, then ξ and

$e\xi$ are said to be associates. An element of I $(\sqrt{m})$ is called a prime element if it is divisible by units or by their associates and not by any other element. If $\xi = \gamma + \delta\sqrt{m}$ is an algebraic number, then $\gamma - \delta\sqrt{m}$ is called the conjugate of ξ and is written as ξ.

Theorem 50. *If e_1 and e_2 are units, then $e_1 . e_2$ and $\frac{e_1}{e_2}$ are also units.*

Proof. We can choose two algebraic integers δ_1 and δ_2 such that $e_1\delta_1 = 1$ and $e_2\delta_2 = 1$.

So, $\qquad e_1e_2\delta_1\delta_2 = 1.$

So, e_1e_2 is a unit. Again, since $e_1\delta_1 = 1 = e_1\delta_2$, each of δ_1 and δ_2 is a unit. So, $\frac{1}{e_1}$ and $\frac{1}{e_2}$ are units. Now since e_1 and $\frac{1}{e_2}$ are units,

$e_2 . \frac{1}{e_2} - \frac{e_1}{e_2}$ is a unit and this completes the proof.

If $\qquad \xi = u + v\sqrt{m}$, then

$$N(\xi) = \xi\bar{\xi} = (u + v\sqrt{m})(u - v\sqrt{m}) = u^2 - mv^2$$

will be called the norm of ξ. If $\xi = a$ (an integer), then $N(\xi) = a^2$, If $m = 2$ or 3 (mod 4), then any element of $I(\sqrt{m})$ can be written as $\xi = a + b\sqrt{m}$ and hence, $N(\xi)$ can be written as:

$$N(\xi) = a^2 - mb^2 \; [m \equiv 2 \text{ or } 3 \pmod 4],$$

and this is an integer.

If $m \equiv 1$ (mod 4), and if ξ is in $I(\sqrt{m})$, then

$$\xi = a + bp = +\frac{1}{2}b\,2(\sqrt{n-1})$$

and $\qquad N(\xi) = (a - \frac{1}{2}b)^2 - \frac{1}{4}mb^2 \; [m \equiv 1 \pmod 4]$

This is also an integer, because 4 | (m – 1) and

$$N(\xi) = (a - \frac{1}{2}b)^2 - \frac{1}{4}mb^2$$

$$= a^2 - ab - \frac{1}{4}(m-1)b^2$$

If $\xi = a + b\sqrt{m}$ and $\eta = c + d\sqrt{m}$, then

$$N(\xi\eta) = \{a + b\sqrt{m}\}\{c + d\sqrt{m}\}\{a - b\sqrt{m}\}\{c - d\sqrt{m}\}$$

$$= (a^2 - mb^2)(c^2 - md^2)$$

$$= N(\xi)\,N(\eta)$$

Theorem 51. *The norm of a unit is +1 or –1 and the algebraic integer whose norm is +1 or –1 is a unit.*

Proof. If e is a unit, then for some algebraic integer δ, $e\delta = 1$ and

$$N(e)\,N(\delta) = N(e\delta) = N(1) = 1 \text{ so,}$$

$$N(e) = +1 \text{ or } -1.$$

If $N(\xi) = +1$ or -1, then $\xi\bar{\xi} = +1$ or -1 So, ξ divides 1 and ξ is a unit.

If $m > 1$, then $N(\xi) = a^2 - mb^2 = 1$

has infinitely many solutions and this will be discussed in the subsequent chapters.

If $m < 0$, by writing $m = -\mu$, we get

$$a^2 - \mu b^2 = 1 \;[m \equiv 2, 3 \pmod 4]$$

$$(a - \frac{1}{2}b)^2 + \frac{1}{4}\mu b^2 = 1 \;[m \equiv 1 \pmod 4],$$

and these equations are true only for finite number of a and b.

If $\mu \geq 4$, then $a = \pm 1$ and $b = 0$.

If $\mu = 1$, then $m \equiv 3 \pmod 4$ and

$a = 0$, $b = \pm 1$ or $a = \pm 1$, $b = 0$

If $\mu = 2$, $m \equiv 2 \pmod 4$ and $a = \pm 1$, $b = 0$

If $\mu = 3$, then $m = 1 \pmod 4$ and

$$a = \pm 1,\ b = 0 \text{ or } a = 0,\ b = \pm 1 \text{ or, } a = b = \pm 1$$

If $m > 0$, since $N(\xi) = a^2 - mb^2$, this may be negative.

Let $M(\xi) = \{N(\xi)$ if $N(\xi) > 0$,

$-N(\xi)$ if $N(\xi) < 0$.

Now, using $M(\xi)$ instead of $N(\xi)$ and arguing as in Theorem 36, we can prove

Theorem. *The algebraic integers whose norms are prime numbers are themselves prime elements (prime algebraic integers). Similarly, using $M(\xi)$ instead of $N(\xi)$ and adapting the arguments of Theorem 38 we get*

Theorem 52. *Each non-zero algebraic integer which is not a unit, is a product of prime algebraic integers.*

But this representation of some algebraic integer may not be unique. For example,

(1) Let $m = -5 \equiv 3 \pmod 4$. The algebraic integers of $K(\sqrt{-5})$ can be expressed as $a + b\sqrt{-5}$. In this case, if

$$1 + \sqrt{-5} = (a + b\sqrt{-5})(c + d\sqrt{-5}), \text{ then}$$

$$6 = (a^2 + 5b^2)(c^2 + 5d^2)$$

So, $a^2 + 5b^2 = 1$, 2 or 3. But it is impossible that $a^2 + 5b^2 = 2$ or 3.

So, $1 + \sqrt{-5}$ is a prime element of $I(\sqrt{-5})$. Similarly if

$$3 = (a + b\sqrt{-5}).(c + d\sqrt{-5}), \text{ then}$$

$$9 = (a^2 + 5b^2)(c^2 + 5d^2)$$

So, $a^2 + 5b^2 = 1$ or 3 and hence 3 is a prime element of $I(\sqrt{-5})$.

Similarly, we can prove that 2 and $(1-\sqrt{-5})$ are prime elements of $I(\sqrt{-5})$. But

$$6 = 2.3 = (1+\sqrt{-5})(1-\sqrt{-5})$$

Thus, 6 can be represented as a product of prime elements of $I(\sqrt{-5})$, in two different ways.

(2) Let $m = 10 \equiv 2 \pmod 4$. Each element of $I(\sqrt{10})$ can be expressed as $a+b\sqrt{10}$. If

$$2 = (a+b\sqrt{10})(c+d\sqrt{10}), \text{ then}$$

$$4 = (a^2 - 10b^2)(c^2 - 10d^2)$$

So, $\quad a^2 - 10b^2 = \pm 2$

or $\quad a^2 \equiv \pm 2 \pmod{10}$

or $\quad a^2 \equiv \pm 2 \pmod 5$

But $\frac{2}{5} \equiv 2^{\frac{1}{2}(5-1)} \equiv 4 \equiv -1 \pmod 5$. So, 2 is a quadratic non-residue mod 5. Similarly, -2 is a quadratic non-residue mod 5. Therefore, it is impossible that

$$a^2 \equiv \pm 2 \pmod 5$$

and it is impossible that

$$2 = (a+b\sqrt{-5})(c+d\sqrt{-5})$$

Similarly it can be proved that each of 4, $4+\sqrt{10}, 4-\sqrt{10}$ is a prime element. But

$$6 = 2.3 = (4+\sqrt{10})(4-\sqrt{10})$$

Thus, the elements of $I(\sqrt{10})$ cannot be expressed uniquely as the product of prime elements.

5.5. We now determine prime elements of Z + *i*Z. Let *p* be a prime number and let it be divisible by $a + ib$. If

$$\pi\lambda = p \text{ then } N(\pi)\, N(\lambda) = N(p) = p^2$$

So, either $N(\lambda) = 1$ or $N(\lambda) = p$. If $N(\lambda) = 1$, then λ is a unit and π is an associate of *p*. If $N(\lambda) = p$, then $N(\pi) = a^2 + b^2 = p$

(*i*) If $p = 2$, then $p = 1^2 + 1^2 = (1 + i)(1 - i) = i(1 - i)^2$. So $1 + i, 1 - i, -1 + i, -1 - i$ are prime elements of Z + *i*Z and these are associates of each other.

(*ii*) Let $p = 4n + 3$, then if $p = a^2 + b^2$, one of *a* and *b* is odd and the other is even. So,

$$p = a^2 + b^2 \equiv 1 \pmod 4$$

But this is impossible. So, prime numbers which can be written as $(4n + 3)$ cannot be expressed as $p = a^2 + b^2$. Each such *p* is a prime element of Z + *i*Z.

(*iii*) If $p = 4n + 1$, then since Legendre symbol of -1 is

$$\left(\frac{-1}{p}\right) = (-1)^{\frac{1}{2}(p-1)} = 1, \quad -1 \text{ is a quadratic residue.}$$

So, $$x^2 \equiv -1 \pmod p$$

has a solution. If *x* is a solution, then $p \mid (x^2 + 1)$. So, $p \mid (x + i)(x - i)$. But *p* neither divides $(x + i)$ nor does it divide $(x - i)$. Therefore, *p* is not a prime element of Z + *i*Z. So, $p = \pi\lambda$ and if $\pi = a + ib$, then $\lambda = a - ib$.

Again,

$$N(\pi) = a^2 + b^2 = p$$

and thus *p* is a sum of two squares. In this case, since π is a prime number, λ is also a prime number because, if

$$\lambda = \delta\eta = (k - il)(u - iv)$$

then $$\pi = (k + il)(u + iv).$$

Hence, π and λ are both prime numbers and each of

$$\pi,\ i\pi,\ -\pi,\ -i\pi,\ \lambda,\ i\lambda,\ -\lambda,\ -i\lambda$$

divides *p*. If $p = c^2 + d^2$, then $(c + id)$ divides *p*, and hence $(c + id)$

is equal to any one of the above eight elements. Since all the eight elements written above are expressible in the form

$$\pm a \pm ib \quad \text{or} \quad \pm ia \pm b,$$

$$c + id = \pm a \pm ib,\ c^2 = a^2,\ d^2 = b^2$$

or, $$c + id = \pm b \pm ia,\ c^2 = b^2,\ d^2 = a^2$$

Therefore,

Theorem 53. *Any prime number which can be written in the form of* $(4n + 1)$ *can be expressed uniquely as the sum of two squares.* *Again*

Theorem 54. *The following are the only prime elements of* $Z + iZ$:

(*i*) *(1 + i) and its associates,*

(*ii*) *any prime number which can be expressed in the form of* $(4n + 3)$,

(*iii*) *the factors of the form* $(a + ib)$ $(b \neq 0)$ *of prime numbers which can be expressed in the form* $(4n + 1)$.

5.6. In this section, we determine the prime elements of $Z + wZ$. We write p for any prime number.

1. If $p = 3$, then $p = (1 - w)(1 - w^2) = (1 + w)(1 - w^2) = -w^2(1 - w)$ and $N(1 - w) = 3$. Therefore, $(1 - w)$ is a prime algebraic integer.

2. Let $p \equiv 2 \pmod 3$. If π is a prime algebraic element, then it is not true that $N(\pi) = \pi$. Because if $\pi = a + bw$, then

$$4N(\pi) = (2a - b)^2 + 3b^2 \equiv (2a - b)^2 \pmod 3 \text{ and}$$

$$(2a - b) \equiv 0, +1 \text{ or } -1 \pmod 3.$$

So, $4N(\pi) \equiv 0$ or $1 \pmod 3$. So, p is not divisible by any algebraic number. Therefore, p is a prime element of $I(\sqrt{-3})$.

3. If $p \equiv 1 \pmod 3$, then from section 2.5, it follows that -3 is a quadratic residue mod 3. Because in this case, $p = 3n + 1$ and hence,

$$x^2 \equiv -3 \pmod p$$

has a solution x. For this x, p divides $(x^2 + 3) = (x + \sqrt{-3})(x - \sqrt{-3})$.

Let p be a prime element of $I(\sqrt{-3})$. Then since $I(\sqrt{-3})$ is Euclidean, p divides $(x+\sqrt{-3})$ or $(x-\sqrt{-3})$.

But this is impossible. So, p is divisible by some prime algebraic integer $(a + bw)$

and $$p = N(a + bw)a^2 - ab + b^2.$$

We have thus proved that

Theorem 55. *The following are the only prime elements of* $Z + wZ$.

(*i*) $1 - w$

(*ii*) *any prime number of the form* $(3n + 2)$, *and*

(*iii*) *the factors of the form* $(a + bw)$ *of the prime numbers which can be expressed in the form* $(3n + 1)$ *and their associates.*

CHAPTER 6

Diophantine Equations

6.1. Let x and y be two adjacent sides of a right-angled triangle and let z be its hypotenuse. Then $x^2 + y^2 = z^2$. Let there be three cubes whose lengths of the sides are x, y and z respectively. If the sum of the weights of the first two cubes is same as that of the third cube, then $x^3 + y^3 = z^3$. In these cases, in order to measure x, y and z, these must be rational numbers. From ancient times, people are trying to solve such equations and till now there is no remarkable success in this direction.

In third century AD, the famous Greece mathematician Diophantus discussed integer solutions of several equations with integer coefficients. Later, such equations are termed Diophantine Equations and the solutions of such equations must be integers or rationals.

6.2. In the seventeenth century AD, the French mathematician *Fermat* conjectured that if $n > 2$, then $x^n + y^n = z^n$ cannot have integer solutions. This conjecture has not yet been established for each n. In this chapter, we shall prove this conjecture for $n = 4$. If $n = 2$, then $x^n + y^n = z^n$ has many integer solutions, for example,

$$(x, y, z) = (3, 4, 5) \text{ or } (5, 2, 13) \text{ or } (8, 15, 17), \ldots$$

We now prove.

Theorem 56. *If x, y and z are positive integers, $(x, y) = 1$ and x is even, then the solution of $x^2 + y^2 = z^2$ is given by $x = 2ab$, $y = a^2 - b^2$, $z = a^2 + b^2$ ($a > b > 0$, $(a, b) = 1$).*

Proof. Since $x^2 + y^2 - z^2 = 0$, every common factor of z and y divides x and every common factor of z and x divides y. So, if $(z, y) = d$, then $(x, y) = d$ and if $(z, x) = d'$; $(x, y) = d'$; Since $(x, y) = 1$, $(y, z) = 1 = (z, x)$. Thus, x, y and z are coprime.

Since x is even, y and z must be odd and $\left(\frac{z+y}{2}\right), \left(\frac{z-y}{2}\right)$ are two integers coprime to each other. But

$$\left(\frac{x}{2}\right)^2 = \left(\frac{z+y}{2}\right)\left(\frac{z-y}{2}\right)$$

So, each of $\left(\frac{z+y}{2}\right)$ and $\left(\frac{z-y}{2}\right)$ is a square. Thus, for some integers a and b, if

$$\left(\frac{z+y}{2}\right) = a^2, \left(\frac{z-y}{2}\right) = b^2, \text{ then}$$

$(a, b) = 1$ and $\frac{x}{2} = ab$. Note that between a and b, one is even and the other is odd. Because if both are odd, ($a^2 \equiv 1 \equiv b^2 \pmod 4$ or $z = a^2 + b^2 \equiv 2 \pmod 4$ and z is even). Again,

$$x^2 + y^2 = (2ab)^2 + (a^2 - b^2)^2 = a^4 + b^4 + 2a^2b^2$$
$$= (a^2 + b^2)^2 = z^2$$

If y and z are given, then $a = \frac{z+y}{2}$ and $b = \frac{z-y}{2}$. Clearly, if a and b are given, then z and y are uniquely determined.

Finally if $(X, Y) = d$ and $X^2 + Y^2 = Z^2$, then

clearly, $(X, Y, Z) = d$ and the solutions of the last equation are given by

$$[X, Y, Z] = [2abd, (a^2 - b^2)d, (a^2 + b^2)d], (a, b) = 1.$$

6.3. $x^4 + y^4 = z^4$ is the Fermat's equation for $n = 4$ and we shall show that this has no integer solution. We shall, in fact, show that if

$$xyz \neq 0,$$
$$x^4 + y^4 = z^2$$

has no integer solution and we do this with the help of *Fermat's Method of Descent.* Let u be the smallest of all z for which (x, y, z) is a solution of $x^4 + y^4 = z^2$ and x, y, z are positive. We will show that if $x^4 + y^4 = z^2$ has an integer solution, then we can find positive integers r, s and d such that $d < u$ and $r^4 + s^4 = d^2$. This would contradict the definition of u and hence $x^4 + y^4 = z^2$ would not have a solution. Again if $(x, y) = d$, then if $d > 1$ by writing $x = d.X$, $y = d.Y$, we observe that d^4 divides $x^4 + y^4 = z^2$. Thus, d^2 / z. We can divide x and y by d and take $(x, y) = 1$. Thus, for every (positive) integer solution (X_1, Y_1, Z_1) of $x^4 + y^4 = z^2$ there is a (positive) integer solution (X_1, Y_1, Z_1) of $x^4 + y^4 = z^2$. Therefore, in order to prove the assertion, it is enough to show that

$$X^4 + Y^4 = Z^2,\ (X, Y) = 1$$

does not have any (positive) integer solution.

Proof. Let $\mho$ be the smallest value of Z for all solutions X, Y, Z of

$$X^4 + Y^4 = Z^2\ (X > 0, Y > 0, Z > 0, (X, Y) = 1.$$

Since $(X, Y) = 1$, at least one of X and Y must be odd. So, $\mho^2 = X^4 + Y^4 \equiv 1$ or $2 \pmod 4$. But $\mho^2 \equiv 2 (\text{mod } 4)$ is impossible. So, $\mho^2 \equiv 1(\text{mod } 4)$, thus, $\mho$ is odd and X or Y is odd.

Let X be even and $\mho$ and Y odd. According to the discussions of the previous section,

$$X^2 = 2ab,\ Y^2 = a^2 - b^2,\ \mho = a^2 + b^2,\ (a, b) = 1$$

In this case, one of a and b is even and other odd. Firstly, if a is even and b is odd, then $b^2 \equiv -1$. But this is impossible. Hence a is odd and b is even. If we take $b = 2c$, then

$(a, c) = 1$ and $\left(\frac{1}{2}X\right)^2 = ac$. So, $a = d^2$, $c = f^2$, $d > 0$, $f > 0$, $(d, f) = 1$ and d is even.

Hence $\quad Y^2 = a^2 - b^2 = d^4 - 4f^4$, *i.e.*, $(2f^2)^2 + Y^2 = (d^2)^2$

and $\quad (2f^2, y) = (y, d^2) = (d^2, 2f^2) = 1.$

Hence, as per the discussions of the previous sections for positive integers l and m which are coprime, we have

$$2f^2 = 2lm \text{ and } d^2 = l^2 + m^2$$

Again, $\quad f^2 = lm$ and $(l, m) = 1$

and hence $\quad l = r^2,\ m = s^2$ and $(l, m) = (r^2, s^2) = 1$

So, $\quad r^4 + s^4 = d^2,\ d \le d^2 = a \le a^2 < a^2 + b^2 = u$

and consequently, there is a positive Z less than Ʊ for which $X^4 + Y^4 = Z^2$ has a solution. But this violates the definition of Ʊ. Therefore, in this case and consequently, in every case

$$X^4 + Y^4 = Z^2 \text{ does not have a solution.}$$

6.4. Pell's Equation

Let D be a positive integer which is not a square. Then

$y^2 - Dx^2 = 1$ has infinitely many different solutions. Among all such solutions if Ʊ is the smallest positive value of x and if $(x, y) = (Ʊ, T)$, then all the solutions can be obtained from

$$y + x\sqrt{D} = \pm(T + Ʊ\sqrt{D})^n \quad (n = \pm1, \pm2, \pm3, \ldots).$$

We need two lemmas for the proof of this.

Lemma 1. If θ is irrational and q is a positive integer, then we can obtain two integers x and y such that

$$L = y - x\theta \text{ and } |\,L\,| < \frac{1}{q}.$$

Proof. For each $x = 1, 2, 3, \ldots, (q + 1)$, choose an integer y such that

$$y - 1 < x\,\theta < y$$

In this case, $x\theta$ is not an integer. Because if $x\theta = z$ (an integer), then

$\theta = \dfrac{z}{x}$ is a rational number, For each x, we have a unique y and if $L = y - x\theta$, then $0 < L < 1$.

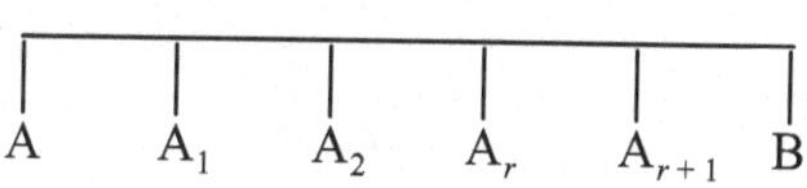

Let AB be a straight line of length l. Divide AB into q equal parts by the points $A_0 = A, A_1, A_2, \ldots, A_q = B$. The length of AA_s, will be $\left(\dfrac{s}{q}\right)$. Suppose that when $x = K$, $y = y_k$ and $L = L_k = y_k - k\theta$. For

$K = 1, 2, 3,..., (q + 1)$, the number L_k will be represented by some point on AB which is different from $\{A_1, A_2,...,A_q\}$. For if L_k will be represented by the point A_r, then

$$L_k = \left(\frac{r}{q}\right)$$

and so,
$$\theta = \frac{1}{K}\left(y_k - \frac{r}{q}\right)$$

which is a rational number. Thus, every L_k represents some points in between some segement A_s, A_{s+1}.

Suppose that none of the segments AA_1, A_1A_2,..., A_{a-1} B contain more than one L_k, then the total number of such points cannot be more than q. But there are $(q + 1)$ such points on AB. Therefore, there are two points L_g and L_h in some segements A_s and A_{s+1}.

So,
$$\frac{s}{q} < y_g - g\theta < \frac{s+1}{q}$$

and
$$\frac{s}{q} < y_h - h\theta < \frac{s+1}{q}$$

Therefore, if $g < h$

$$-\frac{1}{q} < (y_h - y_g) - (h - g)\theta < \frac{1}{q}$$

If we take $y = y_h - y_g$ and $x = h - g$, then

$$| L | = | y - x\theta | = \left| (y_h - y_g) - (h - g)\theta \right| < \frac{1}{q}.$$

Since
$$0 < x = h - g < q,\ \frac{1}{q} < \frac{1}{x},$$

So
$$| y - x\theta | < \frac{1}{q} < \frac{1}{x}$$

Suppose that for $x = x_1$ and $y = y_1$,

$$| y_1 - x_1\theta | < \frac{1}{x_1}.$$

Choose an integer q_1 such that

$$| y_1 - x_1\theta | < \frac{1}{q_1}$$

For the q_1, there are x_2 and y_2 such that

$$| y_2 - x_2\theta | < \frac{1}{q_1} < \frac{1}{x_2}.$$

As before we can choose an integer q_2 such that

$$| y_2 - x_2\theta | < \frac{1}{q_2}$$

and for this q_2 there are x_3 and y_3 such that

$$| y_3 - x_3\theta | < \frac{1}{q_2} < \frac{1}{x_3}.$$

Continuing in this way, we can get an infinitely many $x_1, x_2, \ldots,$ such that for $y = y_k$ and $x = x_k$,

$$| y - x\theta | < \frac{1}{x}.$$

In this case, since,

$$\frac{1}{q_k} < | y_k - x_k\theta | < \frac{1}{q_{k-1}}$$

and

$$\frac{1}{q_{k-1}} < | y_{k-1} - x_{k+1}\theta | < \frac{1}{q_k},$$

(x_{k+1}, y_{k+1}) is different from (x_k, y_k). We thus proved that for infinitely many different (x, y)

$$| y - x\theta | < \frac{1}{x}.$$

Lemma 2. For each D, we can obtain a real number M(D) such that for infinitely many x, y,

$$| y^2 - Dx^2 | < M(D).$$

Proof. Take $\theta = \sqrt{D}$ in the previous lemma. We can get x, y such that

$$| y - x\sqrt{D} | < \frac{1}{|x|}$$

Then $| y + x\sqrt{D} | = | y - x\sqrt{D} + 2x\sqrt{D} |$

$$< | y - x\sqrt{D} | + 2 | x\sqrt{D} |$$

$$< \frac{1}{|x|} + 2|x|\sqrt{D}$$

and hence,

$$| y^2 - x^2D | < \frac{1}{x^2} + 2\sqrt{D} < 2\sqrt{D} + 1 = M(D),$$

Since $\left| y - x\sqrt{D} \right| < \frac{1}{|x|}$ is true for infinitely many (x, y), for these (x, y), we have

$$| y^2 - Dx^2 | < M(D)$$

Since D is an integer, $| y^2 - Dx^2 |$ has infinitely many values which are all positive integers. Let $N \leq D < N + 1$.

If for finitely many (x, y), $| y^2 - Dx^2 | = 1$, $| y^2 - Dx^2 | = 2, \ldots$, $| y^2 - Dx^2 | = N$ and since for all such (x, y),

$$| y^2 - Dx^2 | < M(D),$$

it follows that the above inequation has only finite number of solutions.

But this is false. Therefore, for some positive integer $| k | < M(D)$,

$$y^2 - Dx^2 = k$$

has infinitely many solutions (x, y). Let us denote solutions by $(x_1, y_1), (x_2, y_2),..., (x_n, y_n),....$ Assume that each (x_m, y_m) is positive and

$$x_n = q_n k + r_n, \; - k < r_n < k$$

and $$y_n = q_n' k + r_n', \; - k < r_n' < k$$

In this case the number of $(r_1, r_1') (r_2, r_2'),..., (r_n, r_n')$, ... cannot exceed k^2. So, two of them, say, (r_1, r_1') and (r_2, r_2') must be equal and for this

$$(x_2, y_2) \neq (-x_1, -y_1)$$

So, $x_2 \equiv x_1 \pmod{k}$, and $y_2 \equiv y_1 \pmod{k}$ and $(x_2, y_2) \neq (-x_1, -y_1)$. Multiplying

$$y_1^2 - Dx_1^2 = k$$

and $y_2^2 - Dx_2^2 = k$, we get

$$(y_1y_2 - Dx_1x_2)^2 - D(y_1x_2 - Dy_2x_1)^2 = k^2$$

Write $y_1y_2 - Dx_1x_2 = kY$ and

$$y_1x_2 - y_2x_1 = kX$$

Clearly, X and Y are integers, $X \neq 0$,

and $Y^2 - DX^2 = 1$

If we write –X for X and –Y for Y, then the equation is not unchanged. Therefore, we may assume that X and Y are positive.

Let (X, Y) = (Ʊ, T), Ʊ > 0, T > 0 and Ʊ be the smallest of all the values of X. Then, for n = 1, 2, 3,... and positive x_n, y_n, the infinitely many solutions (x_n, y_n) are obtained from the following relations:

$$y_n + x_n \sqrt{D} = (T + Ʊ\sqrt{D})^n$$

and $$y_n - x_n \sqrt{D} = (T + Ʊ\sqrt{D})^n$$

Again, all the solutions are obtained from the above relations.

Proof. Suppose that (x, y) is a solution which is not obtained from the above relations (Here $x > 0$, $y > 0$). Then, for some positive integer n,

$$(T+\mho\sqrt{D})^n < y + x\sqrt{D} < (T+\mho\sqrt{D})^{n+1}, \text{ and hence,}$$

$$1 < (y + x\sqrt{D})\,(y - x\sqrt{D}) < T+\mho\sqrt{D}$$

In this case, $(y + x\sqrt{D})\,(y_n - x_n\sqrt{D})$

$$= (yy_n - xx_n D) + (xy_n + yx_n)\sqrt{D}$$

$$= Y + X\sqrt{D}$$

So, $Y + X\sqrt{D} < T+\mho\sqrt{D}$

But $y^2 - DX^2 = 1$

and $Y + X\sqrt{D} > 1$

So, $0 < Y - X\sqrt{D} \le 1$

So, $2X\sqrt{D} > 0$, or, $X > 0$

Hence $Y > X\sqrt{D} > 0.$

Again, since $Y^2 - DX^2 = T^2 - D\mho^2 = 1$, we have

$$Y - X\sqrt{D} > T-\mho\sqrt{D}$$

i.e., $-Y + X\sqrt{D} < -T+\mho\sqrt{D}$

Adding this with

$$Y + X\sqrt{D} < T+\mho\sqrt{D}, \text{ we get}$$

$$2x\sqrt{D} < 2\mho\sqrt{D}$$

So, $0 < X < \mho$. But this is impossible. So, there is no solution (x, y) which is not obtained from

$$x + y\sqrt{D} = (T+\mho\sqrt{D})^n$$

and $x - y\sqrt{D} = (T-\mho\sqrt{D})^n$

Corollary. If d is a given positive integer, then for infinitely many solutions, $x \equiv 0 \pmod{d}$.

This is clear from the equation $Y^2 - Dd^2 X^2 = 1$.

If the solutions of this equation are (ξ_k, n_k). $k = 1, 2, \ldots$, then the solutions of $y^2 - Dx^2 = 1$ are $(d\xi_k, n_k)$.

Application. Let $m \equiv 2$ or $3 \pmod{4}$. Then, the algebraic integers of $I(\sqrt{m})$ can be expressed as $(a + b\sqrt{m})$, where a and b are integers.

If $(x + y\sqrt{m})$ is a unit, then

$$x^2 - my^2 = 1$$

or,

$$x^2 - my^2 = -1$$

Similarly, if $m \equiv 1 \pmod{4}$, then the units of $I(\sqrt{m})$ can be written as $(x + \tfrac{1}{2}y\sqrt{m})$ and their norm being ± 1, we have

$$(x + \tfrac{1}{2}y)^2 - (\tfrac{1}{2}my)^2 = 1$$

or

$$(x + \tfrac{1}{2}y)^2 - (\tfrac{1}{2}my)^2 = -1$$

So, in order to determine units, we have to solve

$$x^2 - my^2 = 4$$

or

$$x^2 - my^2 = -4$$

and obtain the values of x and y.

Let $(x, y) = (u, t)$, $u > 0$, $t > 0$ be the solution of

$$y^2 - mx^2 = 4$$

where x is the smallest. Then, all the solutions of the above equation can be obtained from

$$\left(\frac{y + x\sqrt{m}}{2}\right) = e\left(\frac{t + u\sqrt{m}}{2}\right)^n$$

and

$$\frac{y - x\sqrt{m}}{2} = e\left(\frac{t - u\sqrt{m}}{2}\right)^n$$

where $e = \pm 1$ and $n = 1, 2, 3\ldots$

It is difficult to solve

$y^2 - mx^2 = -1$ and we shall discuss its solution when m is a prime number p and $m \equiv 1 \pmod 4$.

Let $(x, y) = (\mho, \mathrm{T})$, $\mho > 0$, $\mathrm{T} > 0$ be a solution of $y^2 - px^2 = 1$ and for all other positive solutions (x, y) of this equation, $x = \mho$. In this case, $\mho$ is even and T is odd, because $p \equiv 1 \pmod 4$. If $\mho \equiv 1 \pmod 2$ and $\mathrm{T} \equiv 0 \pmod 2$, by taking $\mathrm{T} = 2t$, $p = 4q + 1$, $\mho = 2u + 1$, we observe that $\mathrm{T}^2 - p\mho^2 = (2t)^2 - (4q+1)(2u+1)^2 = 4\nu - 1$. But there is no integer ν such that $4\nu - 1 = 1$. This shows that $\mho$ is even and T is odd. So, $\dfrac{\mathrm{T}+1}{2}, \dfrac{\mathrm{T}-1}{2}$ and $\dfrac{\mho}{2}$ are integers

and since $\left(\dfrac{\mathrm{T}+1}{2}\right)\left(\dfrac{\mathrm{T}-1}{2}\right) = p\left(\dfrac{\mho}{2}\right)^2$, for integers a and b,

(*i*) $\dfrac{\mathrm{T}+1}{2} = pa^2, \dfrac{\mathrm{T}-1}{2} = b^2$

(*ii*) $\dfrac{\mathrm{T}+1}{2} = a^2, \dfrac{\mathrm{T}-1}{2} = pb^2$

If (*ii*) holds, then since $\left(\dfrac{\mho}{2}\right)^2 = (ab)^2$, we have

$$a^2 - pb^2 = 1$$

and $(\mho, \mathrm{T})$ will not be the solution of

$$y^2 - px^2 = 1$$

with the smallest positive x. So, (*i*) must hold and

$$b^2 - pa^2 = -1$$

Then, the infinitely many solutions of

$$y^2 - px^2 = -1$$

will be obtained from

$$y + x\sqrt{p} = (b + a\sqrt{p})^{2n+1}$$

and $$y - x\sqrt{p} = (b - a\sqrt{p})^{2n+1} \quad (n = 1, 2, 3, ...)$$

6.5. Let a, b, c, d, e and f be integers such that

$$D = b^2 - 4ac > 0 \text{ and}$$

$$\Delta = 4acf + bde - ae^2 - cd^2 - fb^2 \neq 0.$$

If for some integer x_0, y_0,

$$ax_0^2 + bx_0y_0 + cy_0^2 + dx_0 + ey_0 + f = 0,$$

and if D is not a square, then

$$ax^2 + bxy + cy^2 + dx + ey + f = 0$$

has infinitely many integer solutions.

Proof. The last equation can be expressed as:

$$Y^2 - DX^2 = M$$

Here, $D > 0$, and for some integers p, q, r, s and t,

$$Y = px + qy + r, \quad X = sy + t$$

Since $\Delta \neq 0$, $M \neq 0$. Let

$$Y_0 = px_0 + qy_0 + r, \quad X_0 = sy_0 + t$$

Then, $$Y_0^2 - DX_0^2 = M$$

Again, by section 6.1,

$$\eta^2 - ps\xi^2 = 1$$

has infinitely many integer solutions (ξ, η). If

$X = \eta X_0 - ps\xi Y_0$, $Y = \eta Y_0 - Dps\xi X_0$, then we have

$$Y^2 - DX^2 = 1$$

Here, $\eta^2 \equiv 1 \pmod{ps}$. If $\eta \neq 1 \pmod{ps}$, then by taking

$$\eta_1 + ps\xi_1\sqrt{D} = (\eta + ps\xi\sqrt{D})^2, \text{ we have}$$

$$\eta_1 = \eta^2 + p^2s^2\xi^2 D \equiv 1 \pmod{ps}$$

$$\eta_2 = 2ps\eta\xi \pmod{ps}$$

We may assume that $\eta \equiv 1 \pmod{ps}$, *i.e.*,

we consider those solutions (ξ, η) of

$$\eta^2 + Dp^2s^2\xi^2 = 1$$

for which ps divides $(\eta - 1)$. These are infinitely many such integer values of (ξ, η). For any fixed (ξ, η), there is an unique integer λ such that $\eta = 1 + \lambda ps$. For that (ξ, η), we have

$$px + qy + r = Y = \eta Y_0 - Dps\xi X_0$$

$$= (1+\lambda ps)(px_0+qy_0+r) - Dps\xi(sy_0+t)$$

and $$sy + t = X = \eta X_0 - ps\xi Y_0$$

$$= (1+\lambda ps)(sy_0+t) - ps\xi(px_0+qy_0+r)$$

So, $$x = x_0 + \lambda(Y_0 s - X_0 q) + \xi(qY_0 - DsX_0)$$

$$y = y_0 + \lambda pX_0 - \xi pY_0$$

Let $(\xi_1, 1 + \lambda_1 ps)$ and $(\xi_2, 1 + \lambda_2 ps)$ be two different solutions of

$$\eta^2 - Dp^2s^2\xi^2 = 1$$

Then, $$x_1 - x_2 = (\lambda_1-\lambda_2)(Y_0 s - X_0 q) + (\xi_1-\xi_2)(qY_0 - Dsx_0)$$

$$y_1 - y_2 = (\lambda_1 - \lambda_2)\,pX_0 - (\xi_1 - \xi_2)\,pY_0$$

If $x_1 - x_2 = 0 = y_1 - y_2$, then since

$$\lambda_1 - \lambda_2 \neq 0 \neq \xi_1 - \xi_2$$

$$ps(Y_0^2 - DX_0^2)^2 = 0$$

or $$Y_0^2 - DX_0^2 = 0$$

But this is impossible. So if $\xi_1 \neq \xi_2$ and $\lambda_1 \neq \lambda_2$, then

$x_1 \neq x_2$ and $y_1 \neq y_2$. But

$$\eta^2 - Dp^2s^2\xi^2 = 1$$

has infinitely many solutions (ξ, η).

So, $$ax^2 \times bxy + cy^2 + dx + ey + f = 0$$

has infinitely many distinct integer solutions.

Note. If $ax^2 + bxy + cy^2 + dx + ey + f = 0,$

$$X_1 = (4ac - b^2)y + (2ae - bd),$$

$$Y_1 = 2ax + by + d,$$

$$D = b^2 - 4ac,$$

Then, $$X_1^2 - DY_1^2 = (2ae - bd)^2 + D(4af - d^2) = M_1$$

By multiplying by D and putting

$Y = DY_1$, $X = X_1$ and $M_1D = M$, we get

$$Y^2 - DX^2 = M$$

Here, in each case, the coefficients are integers.

6.6. Let a, b, c and d be integers, $a \neq b$, $c \neq d$, $a - b \equiv 0 \equiv c - d \pmod 2$, $-(a + b)(c + d) > 0$

and $-(a + b)(c + d)$ is not a square, then

$$x^3 + y^3 + z^3 + w^3 = a^3 + b^3 + c^3 + d^3$$

has infinitely many integer solutions (x, y, z, w).

Proof. Put $x = a + X$, $y = b - X$, $z = c + Y$, $w = d - Y$.

Then $(a + b) X^2 + (a^2 - b^2) X + (c + d) Y^2 + (c^2 - d^2) Y = 0.$

Here $D = -4(a + c)(b + d) > 0,$

$$\Delta = -(a + b)(c^2 - d^2)^2 - (c + d)(a^2 - b^2)^2$$

$$= -(a + b)(a + a)\{(c - d)(c^3 - d^2) - (a + b)(a^2 - b^2)\}$$

$$\neq 0.$$

So by section 6.5

$$(a + b) X^2 + (c + d)Y^2 + (a^2 - b^2)X + (c^2 - d^2)Y = 0$$

or $$x^3 + y^3 + z^3 + w^3 = a^3 + b^3 + c^3 + d^3$$

has infinitely many integer solutions.

6.7. In this section, we give another application of the ideas discussed in section 6.5. We now prove that

$$ax^3 + ay^3 + bz^3 = bc^3, \quad abc \neq 0$$

has infinitely many integer solutions besides $\{x + y = 0, z = c\}$.

Proof. Firstly, choose any integer t and put

$$z = c + t(x + y)$$

Now dividing the given equation by $(x + y) \neq 0$, we get

$$a(x^2 - xy + y^2) + 3bc^2t - 3bct^2(x + y) + bt^3(x + y)^3 = 0$$

Take $x + y = u$ and $x - y = v$. Then

$$(-3a)v^2 = (a + 4bt^3)u^2 + 12bct^2u + 12bc^2t$$

We know from section 6.5 that if A, B, C and D are integers, D > 0, D is not a square, $\Delta = D(B^2 - 4AC) \neq 0$ and if

$$Dv^2 = Au^2 + Bu + C$$

has one solution $u = u_0$, $v = v_0$, then this equation has infinitely many integer solutions.

Here, if $t = -abk^2$ (k integer), then

$$(-3a)v^2 = (a + 4bt^3)\, u^2 - (12bct^2)\,(u + 12bc^2t)$$

has $\{u = 0, v = 2bcK\}$ as an integer solution. Again, since $t = -abK^2$,

$$\begin{aligned} AD &= (-3a)\,[a + 4b\,(-abK^2)^3] \\ &= 3a^2(4a^3b^4k^6 - a) \\ &= 3a^2(4a^2b^4k^6 - 1) \end{aligned}$$

is always a positive integer for $K \neq 0$. We can take $(4a^2b^4k^6 - 1)$ different from a square for infinitely many K.

Again,
$$\begin{aligned} \Delta = D(B^2 - 4AC) &= 144\; abc^2t\;(bt^3 + a) \\ &= (-144)a^2b^2c^2(a - a^3b^4k^6) \\ &= 144a^2b^2c^2(a^2b^4k^6 - 1) \end{aligned}$$

and $\Delta \neq 0$ for $k \neq 1$.

So, by section 6.5, since $\{u = 0, v = 2bcK \neq 0\}$ is a solution of

$$(-3a)v^2 = (a + 4bt^3)u^2 + (12bct^2)u + (12bc^2t)$$

This has infinitely many integer solutions.

The solution of the original equation is obtained from

$$x = \frac{1}{2}(u + v)$$

$$y = \frac{1}{2}(u - v)$$

So, in order that x and y are integers.

$$u \equiv v \pmod 2 \text{ must hold.}$$

Let $a = 2^g\alpha$, $g \geq 0$, α is an odd integer. Then, take $t \equiv 0 \pmod{2^g}$ in such a way that it does not violate the conditions imposed on t earlier. This is, clearly, not impossible. So, for u and v, we have

$$-3\alpha v^2 \equiv \alpha u^2 \pmod 2$$

i.e.,
$$u^2 \equiv -3v^2 \equiv v^2 \pmod 2$$

i.e.,
$$u \equiv v \pmod 2$$

So, the original equation will have infinitely many integer solutions (x, y).

6.8. Consider the equation

$$x^2 + y^2 = 2z^2,\ x > 0,\ y > 0,\ z > 0,\ (x, y) = 1.$$

In this case, z must be odd. For if z is even, then x and y must be even. But since $(x, y) = 1$, this is impossible. Therefore if

$$x = X + Y,\ y = X - Y, \text{ then}$$

$$(X, Y) = 1 \text{ and } X^2 + Y^2 = Z^2$$

We now obtain

$$X = a^2 - b^2,\ Y = 2ab, (a, b) = 1$$

and hence,

$$x = a^2 - b^2 + 2ab,\ y = a^2 - b^2 - 2ab,\ z = a^2 + b^2$$

6.9. We obtain the solution of

$$x^2 + y^2 = kx^2$$

by using the properties of Gaussian integers. In this Gaussian domain, ± 1, $\pm i$ are the units and except these, no other algebraic element is a unit.

Again, any prime number p such that $p \equiv 1 \pmod 4$ can be expressed as $x^2 + b^2$ (x, y are integers). But any prime number q such that $q \equiv 3 \pmod 4$ cannot be expressed in the above manner. Furthermore, $2 = i\,(1 - i)^2$ is not a prime element of the Gaussian domain. Thus, the prime elements of the Gaussian domain are the factors $(x + iy)$ and $(x - iy)$ of $p \equiv 1 \pmod 4$, $q \equiv 3 \pmod 4$ and $1 + i$.

We now say that every integer can be expressed uniquely as the product of Gaussian prime numbers, (upto units and upto the order).

Because of this

(1) $x^2 + y^2 = z^2,\ (x, y) = 1,\ z > 0$

can be solved easily as follows:

In this case, $(x + iy)$ and $(x - iy)$ are coprime, for otherwise, the common factor of $(x + iy)$ and $(x - iy)$ will divide $2x$ and $2iy$. Therefore, each of $(x + iy)$ and $(x - iy)$ is a square of Gaussian algebraic number. If $(x + iy) = (a + ib)^2$, then

$$x = a^2 - b^2,\ y = 2ab,\ Z = a^2 + b^2,\ (a, b) = 1$$

(2) Similarly, we can obtain the solution of

$$x^2 + y^2 = pz^2,\ x > 0,\ y > 0,\ z > 0,\ (x, y) = 1$$

for a prime $p \equiv 1 \pmod 4$,

Since $p \equiv 1 \pmod 4$, $p = r^2 + s^2 = (r + is)\,(r - is)$

So, $\quad (r + is)\,(r - is)\,Z^2 = (x + iy)\,(x - iy)$

or, $\quad (r + is)\,(r - is)\,(a + ib)^2\,(a - ib)^2 = (x + iy)\,(x - iy)$

and $\quad x + iy = (r + is)\,(a + ib)^2 = (r + is)\,(a^2 - b^2 + 2iab)$

Thus, the solutions can be written as:

$$x = r\,(a^2 - b^2) - 2sab$$
$$y = s\,(a^2 - b^2) + 2rab$$
$$z = a^2 + b^2$$

(3) Consider

$$4(x^2 + y^2) + z^2 = w^2,\ (x, y, z) = 1,\ x > 0,\ y > 0,\ z > 0$$

In this case, if z is even, the w is even. So, the above equation can be reduced to

$$x^2 + y^2 + z^2 = s^2$$

In the following discussions, we assume that z is odd and hence, w is odd. We have

$$4(x^2 + y^2) = w^2 - z^2 = (w + z)(w - z)$$

Let $w = 2m + 1$ and $z = 2n + 1$.

Then, $w + z = 2(m + n + 1)$, $w - z = 2(m - n) = 2(m + n - 2n)$.

So, between $(m - n)$ and $(m + n + 1)$ one is odd and the other is even, Let $(m - n) = f$ be odd.

Then,
$$\left.\begin{aligned} w - z &= 2f \\ w + z &= \frac{2(x_2 + y_2)}{f} \end{aligned}\right\} (f \text{ odd})$$

If for some prime number $q \equiv 3 \pmod 4$,

$$x \equiv y \equiv 0 \pmod{q^{\alpha}} \quad \text{is true, but}$$

$$x \equiv y \equiv 0 \pmod{q^{\alpha+1}} \text{ is false, then}$$

$$(f, q) = 1 \text{ or } f \equiv 0 \pmod{q^{2\alpha}}$$

For if $f \equiv 0 \pmod{q^{\beta}}$, $f \not\equiv 0 \pmod{q^{\beta+1}}$ and $\beta < 2\alpha$, then

$$w \equiv z \equiv 0 \pmod q \text{ and it is false that } (x, y, z) = 1$$

Let $(x + iy) = (a_1 + ib_1)(a_2 + ib_2) \ldots (a_n + ib_n)$.

be the expression of $(x + iy)$ as the product of Gaussian primes. Since f divides $(x^2 + y^2)$, f will be coprime with a prime number q which can be expressed as $(4n + 3)$ or $(q^{\alpha})^2$ which divides f.

Let
$$x = q_1^{\alpha_1} q_2^{\alpha_2} \ldots q_n^{\alpha_n} x_1$$

$$y = q_1^{\alpha_1} q_2^{\alpha_2} \ldots q_n^{\alpha_n} y_1$$

and (x_1, y_1) does not have any prime factor q which is of the form $(4n + 3)$. Then, the possible factors of f are

$$\left(q_1^{\alpha_1}\right)^2, \left(q_2^{\alpha_2}\right)^2, \ldots \left(q_u^{\alpha_u}\right)^2$$

and all the prime factors of (x_1^2, y_1^2).

Let (x_1^2, y_1^2) be divisible by some prime number p, *i.e.*,

$$x_1^2 \equiv -y_1^2 \pmod{p}$$

If p could be expressed as $4k + 3$, then the prime number (-1) would be a quadratic non-residue of p. Again since the prime number y_1^2 is a quadratic residue, if $y_1 \not\equiv 0 \pmod{p}$, then the product of (-1) and $y_1^2 \neq x_1^2 \not\equiv 0 \pmod{p}$ and x_1^2 will be a quadratic non-residue mod p. But this is impossible. So, (*i*) x_1 and y_1 both will be divisible by p and thus it is impossible that $p = 4k + 3$, or (*ii*) p will not divide x_1 or y_1 and in this case also, it is impossible that $p = 4k + 3$. Thus, every prime factor of $(x_1^2 + y_1^2)$ will be of the form $p = 4k + 1$ and can be expressed as $p = a^2 + b^2$. Hence, the possible factors of f are

(*i*) $\left(q_1^{\alpha_1}\right)^2, \left(q_2^{\alpha_2}\right)^2, \ldots \left(q_u^{\alpha_u}\right)^2$ and

(*ii*) every prime factor of $(x_1^2 + y_1^2)$ which is of the form $p = a^2 + b^2$.

Now, since $(a_k + ib_k)$ is a prime factor of $(x + iy)$,

$$f = (a_1^2 + b_1^2)\eta_1(a_2^2 + b_2^2)\eta_2 \ldots (a_n^2 + b_n^2)\eta_n$$

and each η is either 0 or 1. Therefore, since

$$\begin{aligned}(\alpha^2 - \beta^{2)}(\gamma^2 + \delta^{2)} &= (\alpha + i\beta)(\alpha - i\beta)(\gamma + i\delta)(\gamma - i\delta)\\ &= (\alpha + i\beta)(\gamma + i\delta)(\alpha - i\beta)(\gamma - i\delta)\\ &= [(\alpha\gamma - \beta\delta) + i(\beta\gamma + \alpha\delta)][(\alpha\gamma - \beta\delta) - i(\beta\gamma - \alpha\delta)]\\ &= (\alpha\gamma - \beta\delta)^2 + (\beta\gamma + \alpha\delta)^2\\ f &= Z^2 + W^2\end{aligned}$$

In this case, Z and W are integers.

Let $\qquad Z + iW = (a_1 + ib_1)\eta_1(a_2 + ib_2)\eta_2 \ldots (a_n + ib_n)\eta_n$

Then, for integers X and Y

$$(x + iy) = (X + iY)(Z + iW)$$

i.e.,

$$x = XZ - YW,\ y = YZ + XW$$

$$z = X^2 + Y^2 - Z^2 - W^2$$

$$w = X^2 + Y^2 + Z^2 + W^2$$

are the solutions of the given equation.

(4) Consider

$$x^2 + y^2 = z^2 + w^2, (x, y, z) = 1, x > 0, y > 0, z > 0$$

In this case $(x + z)(x - z) = (w + y)(w - y)$.

If $(x + z, w + y) = d_1$ and $(x - z, w - y) = d_2$, then let

$$(x + z) = d_1\text{X}, (w + y) = d_1\text{Z}, (x - z) = d_2\text{Y}, (w - y) = d_2\text{Z}$$

So $$\text{XY} = \text{ZW}$$

Note that Y and W are coprime and also X and Y are coprime.

So, $$\text{X} = \pm\text{W}, \text{Y} = \pm\text{Z}$$

Thus, $$x + z = d_1\text{X}, \quad x - z = d_2\text{Y}$$

$$w + y = ed_1\text{Y}, \quad w - y = ed_1\text{X} \ (e = \pm 1)$$

Therefore,

$$2x = (d_1\text{X} + d_2\text{Y}), 2z = (d_1\text{X} - d_2\text{Y})$$

$$2w = e(d_1\text{Y} + d_2\text{X}), 2y = e(d_1\text{Y} - d_2\text{X})$$

are the solutions of the given equation. Here d_1 and d_2 are any two integers.

(5) For the solution of

$$x^2 + y^2 + z^2 = w^2, (x, y, z) = 1, x > 0, y > 0, z > 0$$

take $x = t\text{X}, y = t\text{Y}, \text{Z} = t\text{Z} - w$

Then $t^2(\text{X}^2 + \text{Y}^2 + \text{Z}^2) = 2twz$ and

$$t^2(-\text{X}^2 - \text{Y}^2 - \text{Z}^2) = (z + w)^2 - z^2 + w^2 = 2w(z + \text{w}) = 2tz\text{Z}$$

So, $$\frac{x}{2\text{XZ}} = \frac{y}{2\text{YZ}} = \frac{z}{\text{Z}^2 - \text{X}^2 - \text{Y}^2} = \frac{w}{\text{Z}^2 + \text{X}^2 + \text{Y}^2} = \frac{t}{2z} = \frac{1}{\lambda}$$

Therefore, the solution is given by

$$\lambda x = 2\text{XZ}, \lambda y = 2\text{YZ}, \lambda x = \text{Z}^2 - \text{X}^2 - \text{Y}^2, \lambda w = \text{Z}^2 + \text{X}^2 + \text{Y}^2$$

where λ is an integer and chosen in such a way that

$$(x, y, z) = 1 = (\text{X}, \text{Y}, \text{Z})$$

For example, take X = 1, Y = 2, Z = 3, then

$\lambda x = 6$, $\lambda y = 12$, $\lambda z = 4$. Here (X, Y, Z) = 1. So take $\lambda = 2$.

Then, $x = 3$, $y = 6$, $z = 2$, $w = 7$ is a solution of

$$x^2 + y^2 + z^2 = w^2$$

and $$(x, y, z) = 1$$

Similarly if X = 2, Y = 3, Z = 4, then $\lambda x = 16$, $\lambda y = 24$, $\lambda z = 3$. Here, (X, Y, Z) = 1. Take $\lambda = 1$. Then $x = 16$, $y = 24$, $z = 3$, $w = 29$ is a solution of

$$x^2 + y^2 + z^2 = w^2$$

and $$(x, y, z) = 1$$

Since x, y, z *and* w are expressed in terms of X, Y and Z these are called the parameters of x, y, z and w.

6.10. In the previous section, we have observed that Diophantine equations have several solutions. In this section we mention that the equations of degrees higher than two have limited number of solutions or have no solutions.

1. $x^4 - y^4 = z^2$, $(x, y) = 1$
   has no solution other than $x^2 = y^2 = 1$ and $x^2 = 1$, $y = 0$
2. $x^4 + y^4 = 2z^2$, $(x, y) = 1$
   has no solution other than $x^2 = y^2 = 1$.
3. $x^4 - 6x^2y^2 + y^4 = z^4$, $(x, y) = 1$
   has no solution except $x = 0$ or $y = 0$.
4. $x^4 + x^2y^2 + y^4 = z^2$, $(x, y) = 1$
   has no solution except $x^2 = 1$, $y = 0$ and $x = 0$, $y^2 = 1$.
5. $x^4 - x^2y^2 + y^4 = z^2$, $(x, y) = 1$
   has no solution except $x^3 = 1$, $y = 0$, $x = 0$, $y^2 = 1$ and $x^2 = y^2 = 1$.

(**Thue-Siegel Theorem**). Let $n \geq 3$, and $a_0, a_1, \dots, a_n$ be integers and let

$$f(x, y) = a_0x^n + a_1x^{n-1}y + \dots + a_ny^n$$

be a homogeneous polynomial. If $f(x, y)$ cannot be expressed as a product of two homogeneous polynomials $g(x, y)$ and $h(x, y)$, then for any integer m, the equation

$$f(x, y) = m$$

either has no solution or the number of solutions form a finite set.

CHAPTER 7

P-Adic Numbers

7.1. Suppose that for some integer n we want to obtain the solution of

$$x^2 \equiv 2 \pmod{7^n}$$

If $n = 1$, then this congruence has two solutions:

$$x_0 \equiv \pm 3 \pmod{7}$$

Let $n = 2$. We know that if

$$x^2 \equiv 2 \pmod{7^2}, \text{ then}$$

it is necessary that $x^2 \equiv 2 \pmod{7}$. Therefore, if x_1 is a solution of $x^2 \equiv 2 \pmod{7^2}$ we must have $x_1 \equiv x_0 \pmod{7}$.

So, $x_1 = x_0 + 7t_1$. Take $x_0 = 3$. Then, $x_1 = 3 + 7t_1$. Now, since

$$(3 + 7t_1)^2 = 2 \pmod{7^2}, \text{ we have}$$

$$9 + 6.7t_1 + 7^2t_1^2 \equiv 2 \pmod{7^2}$$

i.e., $$7(1 + 6t_1) \equiv 0 \pmod{7^2}$$

i.e., $$(1 + 6t_1) \equiv 0 \pmod{7}$$

Hence, $$t_1 \equiv 1 \pmod{7}$$

Thus, $$x_1 \equiv 3 + 7.1 \pmod{7^2}$$

Similarly, let $x_2 = x_1 + 7^2t_2$ be a solution of

$$x^2 \equiv 2 \pmod{7^3}$$

Here $$(3 + 7.1 + 7^2t_2) \equiv 2 \pmod{7^3}$$

i.e., $9 + 7^2.1^2 + 7^4t_2^2 + 6.7 + 6.7^2t_2 + 2.7^3t_2 \equiv 2 \pmod{7^3}$

i.e., $7\{1 + 7 + 7^3t_2^2 + 6 + 6.7t_2 + 2.7^2t_2\} \equiv 0 \pmod{7^3}$

i.e., $$2.7 + 7^3t_2^2 + 6.7t_2 + 2.7^2t_2 \equiv 0 \pmod{7^2}$$

i.e., $$2 + 7^2t_2^2 + 6t_2 + 2.7t_2 \equiv 0 \pmod 7$$

i.e., $$2 + 6t_2 \equiv 2 \pmod 7$$

or $$t_2 \equiv 2 \pmod 7$$

Hence, $x_2 = 3 + 7.1 + 7^2.2 \pmod 7$

If we proceed in this way, we get a sequence $\{x_0, x_1, x_2 \dots \dots, x_n, \dots\}$ of integers such that

$$x_0 \equiv 3 \pmod 7$$

$$x_n \equiv x_{n-1} \pmod{7^n}$$

$$x_n^2 \equiv 2 \pmod{7^{n+1}}$$

Clearly,

$$x_n \equiv (3 + 7.1 + 7^2.2 + 7^3.6 + \dots + 7^n t_n \pmod{7^n})$$

Definition: Let p be a prime number. If the elements of the sequence $\{x_0, x_1, x_2, \dots, x_{n-1}, x_n, x_{n+1}, \dots\}$ are integers and for each $n \geq 1$, $x_n \equiv x_{n-1} \pmod{p^n}$, then the sequence $\{x_0, x_1, x_2, \dots, x_n \dots\}$. or, briefly $\{x_n\}$, is said to represent a p-adic integer. For two sequences $\{x_n\}$ and $\{x_n^1\}$, if for $n = 0, 1, 2, \dots$

$$x_n \equiv x_n^1 \pmod{p^{n+1}}, \text{ then}$$

$\{x_n\}$ and $\{x_n^1\}$ are said to represent the same p-adic integer. If $\{x_n\}$ represents a p-adic integer α, then we write

$$\{x_n\} \rightarrow \alpha$$

The set of all p-adic integers will be denoted by 0_p.

If x is an integer, then the p-adic integer represented by $\{x, x, \dots\}$ will be denoted by x. For two different integers x and y we will obtain two different p-adic integers, for if they will represent the same p-adic integer, then for each n,

$$x \equiv y \pmod{p^n}$$

and this is possible only if $x = y$. Hence the set of all integers, Z, is a subset of the set of all p-adic integers 0_p,

Let a p-adic integer α be given by a sequence $\{x_n\}$.

Let $\qquad \bar{x}_n \equiv x_n \pmod{p^{n+1}}$,

and $\qquad 0 \le \bar{x}_n < p^{n+1}$

Here, $\qquad \bar{x}_n \equiv x_n \pmod{p^n}$

$$x_n \equiv x_{n-1} \pmod{p^n}$$

and $\qquad x_{n-1} \equiv \bar{x}_{n-1} \pmod{p^n}$

Therefore, $\qquad x_n \equiv \bar{x}_{n-1} \pmod{p^n}$

and hence, the sequence $\{\bar{x}_n\}$ will determine some p-adic integers. Since $x_n \equiv \bar{x}_n \pmod{p^{n+1}}$, this p-adic integer will be equal to that determined by the sequence $\{x_n\}$. For the elements of a sequence $\{x_n\}$, if

$$x_n \equiv x_{n-1} \pmod{p^n}$$

$$0 \le x_n < p^{n+1}, \text{ then}$$

the sequence is said to be a canonical sequence. Each p-adic integer is determined by a single canonical sequence. In other words, if two canonical sequences determine a p-adic integer, the two sequences are equal.

Proof. Let $\{\bar{x}_n\}$ and $\{\bar{y}_n\}$ be two canonical sequences which determine one and only one p-adic integer.

So, $\qquad \bar{y}_n = \bar{x}_n \pmod{p^n}$

$$0 \le \bar{x}_n < p^{n+1},\ 0 \le \bar{y}_n < p^{n+1}$$

If $\qquad \bar{x}_n \ge \bar{y}_n$, then for some integer, $a \ge 0$,

$$\bar{x}_n = \bar{y}_n + ap^{n+1}$$

Since $\qquad \bar{y}_n \ge$ and $0 \le \bar{x}_n < p^{n+1}$,

$$0 \le ap^{n+1} < p^{n+1};$$

and since a is an integer, $a = 0$. Therefore, each p-adic integer can be expressed by one and only one canonical sequence.

Again, since $$\bar{x}_{n+1} = \bar{x}_n \pmod{p^{n+1}},$$

$$\bar{x}_{n+1} = \bar{x}_n + a_{n+1}\, p^{n+1}$$

Further, $$0 \le \bar{x}_{n+1} < p^{n+2},$$

$$0 \le \bar{x}_n < p^{n+1}$$

and a_{n+1} is an integer. If $a_{n+1} \le -1$, then $\bar{x}_{n+1} \le \bar{x}_n = p^{n+1} < 0$. But $\bar{x}_{n+1} \ge 0$ and hence it is impossible that $a_{n+1} \le -1$. So, $a_{n+1} \ge 0$. Again, since $0 \le \bar{x}_{n+1} < p^{n+2}$, we have

$$a_{n+1} < p$$

and hence, $$\bar{x}_{n+1} = \bar{x}_n + a_{n+1}\, p^{n+1}$$

$$= \bar{x}_{n-1} + a_n p^n + a_{n+1}\, p^{n+1} = \ldots$$

$$= a_0 + a_1 p + a_2 p^2 + \ldots + a_n p^n + a_{n+1} p^{n+1}$$

Hence, for each prime number p, the canonical representation of any p-adic integer is given by

$$\alpha = \{a_0,\ a_0 + a_1 p,\ a_0 + a_1 p + a_2 p^2,\ \ldots\},\ 0 \le a < p$$

7.2. Let α and β be two p-adic integers. Let $\{x_n\}$ and $\{y_n\}$ be the canonical sequences which represent α and β respectively. In this case,

$$x_n = x_{n-1} \pmod{p^n},$$

and $$y_n \equiv y_{n-1} \pmod{p^n},$$

Hence, $$x_n + y_n \equiv x_{n-1} + y_{n-1} \pmod{p^n}$$

$$x_n y_n \equiv x_{n-1}\, y_{n-1} \pmod{p^n}$$

Therefore, the sequences $\{x_n + y_n\}$ and $\{x_n y_n\}$ determine two p-adic integers; let us denote them by $\alpha + \beta$ and $\alpha.\ \beta$ respectively. $\alpha + \beta$ and $\alpha.\beta$ are said to be the sum and product of α and β respectively.

Again, for any p-adic integer γ, if $\beta = \alpha\gamma$, then we say that α divides β, Let

$$\overline{1} = \{1, 1, 1, ...\}$$

The p-adic integers which divide $\overline{1}$ are called the p-adic units.

Theorem 57. *Let $\alpha = \{x_0, x_1, ..., x_n ...\}$ be a p-adic integer. α is a p-adic unit if and only if $x \equiv 0 \pmod p$.*

Proof. Let α be a p-adic unit and let β a p-adic integer such that

$$\alpha\beta = \overline{1}.$$

If β is determined by the sequence $\{y_n\}$, then since $\alpha\beta = \overline{1}$, we have

$$x_n y_n \equiv 1 \pmod{p^n}$$

So, $\quad x_0 y_0 \equiv 1 \pmod p$

and hence, $\quad x_0 \not\equiv 0 \pmod p$

Conversely, if $\quad x_0 \not\equiv 0 \pmod p$, then since

$x_n \equiv x_{n-1} \equiv ... \equiv x_2 \equiv x_1 \equiv x_0 \pmod p$, we must have

$$x_n \not\equiv 0 \pmod p:$$

Hence for any n we can find y_n such that

$$x_n y_n = 1 \pmod{p^{n+1}}.$$

Again since $x_n \equiv x_{n-1} \pmod{p^n}$ and $x_n y_n \equiv 1 \equiv x_{n-1} y_{n-1} \pmod{p^n}$, we have

$$x_{n-1} y_n \equiv x_{n-1} y_{n-1} \pmod{p^n}$$

So, $x_{n-1}(y_n - y_{n-1})$ is divisible by p^n.

But $x_{n-1} y_{n-1} \equiv 1 \pmod{p^n}$. Therefore x_{n-1} and p^n are coprime. So, p^n divides $(y_n - y_{n-1})$, *i.e.*,

$$y_n \equiv y_{n-1} \pmod{p^n}$$

So the sequence $\{y_n\}$ determined *a* p-adic integer β. Again $x_n y_n \equiv 1 \pmod{p^{n+1}}$. So, $\alpha\beta = 1$ and hence, α is a p-adic unit.

If for each, $n = 0, 1, 2, ..., x_n \equiv 0 \pmod{p^{n+1}}$, then the p-adic number determined by the sequence $\{x_0, x_1, x_2, ..., x_n, ...\}$ is called zero and denoted by $\overline{0}$.

Theorem 58. *Every p-adic integer α different from zero can be expressed uniquely as*

$$\alpha = p^m e$$

for some p-adic unit e and some integer m.

Proof. If α is equal to some p-adic unit, then for $m = 0$ we have $\alpha = p^m e$. Now assume that α is not a p-adic unit. Let $\{x_n\} \to \alpha$. Since α is not a p-adic unit, by the previous theorem we have $x_0 \equiv 0$ (mod p). Again since $\alpha \neq \bar{0}$, $x_n \equiv 0 \bmod (p^{n+1})$ is not true for all n. Let m be the smallest of all n for which $x_n \not\equiv 0 \pmod{p^{n+1}}$. Hence $x_m \not\equiv 0 \pmod{p^{m+1}}$ and for any $s \geq 0$,

$$x_{m+s} \equiv x_{m-1} \equiv 0 \pmod{p^m}$$

Now, since p^m divides x_{m+s},

$$y_s = \frac{x_{m+s}}{p^m} \text{ is an integer.}$$

Here for all $s \geq 0$,

$$p^m y_s - p^m y_{s-1} = x_{m+s} - x_{m+s-1} \equiv 0 \pmod{p^{m+s}}$$

Hence, $\qquad y_s \equiv y_{s-1} \pmod{p^s}$.

Again since $x_m \not\equiv 0 \pmod{p^{m+1}}$, we have

$$y_0 \equiv \frac{x_m}{p^m} \not\equiv 0 \pmod{p}$$

So, the series $\{y_0, y_1, y_2, ...\}$ determines a p-adic unit e.

Finally, for $s = 0, 1, 2, ...$ since

$$p^m y_s = x_{m+s} \qquad = x_s \pmod{p^{s+1}},$$

the sequences $\{p^m y_s\}$ and $\{x_s\}$ determine one p-adic integer,

So, $\qquad \alpha = \{x_0, x_1, x_2, ...\}$

$$= \{p^m y_0, p^m y_1, ...\}$$

or, $\qquad \alpha = p^m e.$

Assume that for some other p-adic unit e' and integer $k \geq 0$, we have $\alpha = p^k e'$. If $\{Z_s\} \to e'$, then for each $s \geq 0$,

$$p^m y_s = p^k z_s \pmod{p^{s+1}}$$

Since e and e' are p-adic units,

$$y_s \equiv y_0 \not\equiv 0 \pmod{p} \text{ and } z_s \equiv z_0 \not\equiv 0 \pmod{p}.$$

If we take $s = m$ in the above congruence, we get

$$p^m y_m = p^k z_m \pmod{p^{m+1}}$$

Since $x^m \not\equiv 0 \pmod{p^{m+1}}$ and $x_m = p^m y_m$, we must have

$$p^k z_m \not\equiv 0 \pmod{p^{m+1}},$$

Consequently, $k \leq m$. Similarly, we can show that $m \leq k$ and, thus, $k = m$. Again since

$$p^m y_{m+s} \equiv p^k z_{m+s} = p^m z_{m+s} \pmod{p^{m+s+1}}, \text{ we}$$

have

$$y_{m+s} \equiv z_{m+s} \pmod{p^{s+1}}$$

But for each $s \geq 0$,

$$y_{m+s} = y_s \pmod{p^{s+1}}$$

and

$$z_{m+s} = z_s \pmod{p^{s+1}}$$

Hence

$$y_s = z_s \pmod{p^{s+2}}$$

Consequently,

$$\begin{aligned} e &= \{y_0, y_1, ...\} \\ &= \{z_0, z_1, ...\} \\ &= e' \end{aligned}$$

Thus, the representation of α is unique.

Definition: If $\alpha \neq \overline{0}$ and $\alpha = p^m e$ for some p-adic unit e and integer $m \geq 0$, then m is said to be the p-adic value of α and will be denoted by $\nu_p(\alpha)$, *i.e.*, $m = \nu_p(\alpha)$.

If we are concerned with the p-adic values with respect to some fixed prime p, then we write $\nu(\alpha)$ for $\nu_p(\alpha)$. We define $\nu(0) = \infty$.

Remark: (1) If $\{x_n\} = \alpha = p^m e$ and e is a p-adic unit, then x_0, x_1, x_2, ..., x_{m-1} will respectively be divisible by p, p^2, ...p^m; p^m will divide each of x_m, x_{m+1}, ..., but p^{m+1} will divide none of these.

(2) If $\alpha \neq \bar{0}, \beta \neq \bar{0}$, then $\alpha\beta \neq \bar{0}$. For, if $\alpha = p^m e$, $\beta = p^n e'$, then $\alpha\beta = p\ e\ e'$. If $e.e_1 = 1'$, $e'e_1' = \bar{1}$ then $\alpha\beta \neq \bar{0}$ gives that

$$\bar{0} = \alpha\beta e_1 e_1' = p^m e e_1 p^n e' e_1' = p^{m+n}$$

But this is impossible and hence, $\alpha\beta \neq \bar{0}$.

(3) If for p-adic units e and e', $\alpha = p^m e$, $\beta = p^n e'$, then for $m \geq n$,

$$\alpha + \beta = p^m e + p^n e' = p^n(p^{m-n} e + e')$$

So, $\nu(\alpha + \beta) \geq n$. But for $m > n$ if

$$e = \{e_k\},\ e' = \{e'k\}, \text{ then}$$

$$p^{m-n} e_0 + e'0 = e'_0 \not\equiv 0 \pmod{p}.$$

Hence, $\delta = (p^{m-n} e + e')$ is a p-adic unit

and $\alpha + \beta = p^n \delta$

So $\nu(\alpha + \beta) = n.$

Hence, if $\nu(\alpha) = m$, $\nu(\beta) = n$ and $m \geq n$,

$$\nu\ (\alpha\beta) = m + n = \nu\ (\alpha) + \nu\ (\beta),$$

$$\nu(\alpha + \beta) \geq n \text{ for } \nu\ (\alpha) \geq \nu\ (\beta),$$

$$\nu(\alpha + \beta) = n \text{ for } \nu\ (\alpha) > \nu\ (\beta).$$

Let the smaller of $\nu\ (\alpha)$ and $\nu\ (\beta)$ be denoted by min $[\nu\ (\alpha), \nu\ (\beta)]$. Then, the above relations can be expressed as:

$$\nu(\alpha\beta) = \nu(\alpha) + \nu(\beta)$$

$$\nu(\alpha + \beta) \geq \min\ [\nu(\alpha), \nu(\beta)],$$

$$\nu(\alpha + \beta) = \min\ [\nu(\alpha), \nu(\beta)],\ [\nu\ (\alpha) \neq \nu(\beta)].$$

(4) If a p-adic number β divides α, then

$$\nu\ (\alpha) \geq \nu\ (\beta)$$

(5) Let e be a p-adic unit. If $ee' = \overline{1}$, then e' is called the inverse of e. Let e_1 and e_2 be the inverse of e. Then,

$$ee' = \overline{1} = ee_2 \text{ and } e(e_1 - e_2) = \overline{0}$$

So, $$(e_1 - e_2) = \overline{1} \cdot (e_1 - e_2) = e_1.e(e_1 - e_2) = e_1.\ \overline{0} = \overline{0}.$$

Thus $e_1 = e_2$. Hence every p-adic unit has one and only one inverse and the inverse of e is denoted by e^{-1}.

(6) If q is a prime number and $q \neq p$, then $e = \{q, q, ...\}$ is a unit since $q \not\equiv 0 \pmod{p}$. Again since every p-adic integer can be expressed as $\alpha = p^m e'$, so if α is not a unit then it must be divisible by p. So there is one and only one p-adic integer which is prime and this is equal to p.

7.3. For two p-adic integers α and β if $(\alpha - \beta)$ is divisible by a p-adic integer γ, then we write

$$\alpha \equiv \beta \pmod{\gamma}.$$

If $\alpha = p^n e$ and e is a unit, then in order that $\alpha \equiv \beta \pmod{\gamma}$ it is necessary and sufficient that

$$\alpha \equiv \beta \pmod{p^n}$$

Theorem 59. *Any p-adic integer is congruent to an integer mod p.*

Proof. Let α be a p-adic integer determined by the sequence $\{x_n\}$. Then the p-adic integer $(\alpha - x_{n-1})$ will be determined by the sequence

$$\{x_0 - x_{n-1}, x_1 - x_{n-1}, x_2 - x_{n-1}, ..., x_k - x_{n-1}, ...\}$$

In this case, since $x_m = x_{m-1} \pmod{p^m}$ for all m, we have

$$x_m \equiv x_{m-1} \equiv ...x_{n-1} \pmod{p^n} \text{ for } m \geq n$$

Again,

$$\begin{aligned}
x_{n-1} &\equiv x_{n-2} \pmod{p^{n-1}} \\
x_{n-2} &\equiv x_{n-3} \pmod{p^{n-2}} \\
x_{n-3} &\equiv x_{n-4} \pmod{p^{n-3}} \\
&\vdots \\
x_2 &\equiv x_1 \pmod{p^2} \\
x_1 &\equiv x_0 \pmod{p}
\end{aligned}$$

i.e., $$x_m \equiv x_{m+1} \equiv \ldots \equiv x_{n-1} \pmod{p^{m+1}}$$

So,

$x_0 - x_{n-1}, x_1 - x_{n-1}, \ldots, x_{n-2} - x_{n-1}$ are respectively divisible by $p, p^2, \ldots p^n$. Also, each of $x_n - x_{n-1}, x_{n+1} - x_{n-1}, \ldots, x_k - x_{k-1}$ is divisible by p^n and none is divisible by p^{n+1}. So, for some unit e, $(\alpha - x_{n-1}) = p^n e$, *i.e.,*

$$\alpha \equiv x_{n-1} \pmod{p^n}$$

Let $x \equiv y \pmod p$. For some m, we can write

$$x - y \equiv p^m a, \ a \not\equiv 0 \pmod p$$

Here, $x \neq y$. If as integers $x \equiv y \pmod{p^n}$, then $n \leq m$. In this case, as a p-adic integer a is a unit and $x - y = p^m a$.

So, $$\nu_p(x - y) = m$$

So, $$n \leq m = \nu_p(x - y).$$

Since, $$\nu_p(p^n) = n \leq \nu_p(x - y),$$

as a p-adic integer, $(x - y) \equiv \overline{0} \pmod{p^n}$

Again, if as a p-adic integer $x - y \equiv \overline{0} \pmod{p^n}$, for some unit e, $x - y = p^n e$. So

$$x - y \equiv 0 \pmod p, (x - y) \equiv 0 \pmod{p^2}, \ldots, (x - y) \equiv 0 \pmod{p^n}$$

Therefore, in order that two integers will be congruent mod p^n it is necessary and sufficient that as p-adic integers they must be congruent mod p^n.

7.4. In section 7.1 we have obtained the canonical sequence of p-adic integer. If the canonical sequence of a p-adic integer α is given by

$$\alpha = \{a_0, a_0 + a_1 p, a_0 + a_1 p + a_2 p^2, \ldots, a_0 + a_1 p + \ldots + a_m p^m, \},$$

then $0 \leq a_n < p$. If for some unit e, $\alpha = p^k e$, then

$$a_0 \equiv 0 \pmod p$$

$$a_0 + a_1 p \equiv 0 \pmod{p^2}$$

$$a_0 + a_1 p + a_2 p^2 \equiv 0 \pmod{p^2}$$

$$\ldots \ldots \ldots \ldots$$

$$a_0 + a_1 p + \ldots\ldots + a_{k-1} p^{k-1} \equiv 0 \pmod{p^k}$$

and therefore,

$$a_0 \equiv a_1 = a_2 = \ldots\ldots = a_{k-1} = 0$$

Therefore, the canonical sequence for $\alpha = p^k e$ is given by k times

$$\{0, 0\ldots, 0, a_k p^k, a_k p^k + a_{k+1} p^{k+1}, \ldots, a_k p^k + \ldots + \ldots$$
$$\ldots + a_{k+n} p^{k+n}\}$$

i.e., if $\qquad \alpha = p^k e \equiv \{\bar{x}_0, \bar{x}_1, \bar{x}_2, \ldots, \bar{x}_m, \ldots\},$

then $\qquad \bar{x}_0 = \bar{x}_1 = \bar{x}_2 = \ldots = \bar{x}_{k-1} = 0$

and for $n \geq k$, $\bar{x}_n = a_k p^k + \ldots + a_n p^n$, $(0 \leq a_n < p)$

In this way, we determine the canonical sequence of each p-adic integer.

CHAPTER 8

Arithmetical Functions

8.1. In this chapter we shall briefly discuss some of the functions used in Number Theory. In order to study them we shall use some of the formulas and notations from calculus and algebra.

Let $\{n_1, n_2, ...\}$ be a subset of N and let for each n_k, $f(n_k)$ have a definite value. Then, f or $f(n)$ is said to be an arithmetical function of n. In the following sections, we shall discuss some special arithmetical functions which are important in Number Theory.

8.2. The Mobius Function. The Mobius Function μ is defined as follows:

(*i*) $\mu(1) = 1$.

(*ii*) if n has a factor which is a square, then $\mu(n) = 0$,

(*iii*) If $p_1, p_2, ..., p_k$ are distinct prime numbers, then

$$\mu(p_1 \, p_2 ... \, p_k) = (-1)^k.$$

It is clear that if $(m, n) = 1$, then $\mu(mn) = \mu(m)\mu(n)$, *i.e.*, μ is a multiplicative function.

Let $n > 1$ and let n have k distinct prime factors. Let its divisors arranged in an ascending order be $d_1, d_2, ..., d_m$ $(d_1 \leq d_2 < ... < d_m)$. We now prove that

(*i*) $\mu(d_1) + \mu(d_2) + ... + \mu(d_m) = 0$ and,

(*ii*) $|\mu(d_1)| + |\mu(d_2)| + ... + |\mu(d_m)| = 2^k$

Proof. Case I. If n is a prime number p, then $k = 1$, $1 = d_1$, $p = d_2$ and n has only two divisors. So, $\mu(d_1) = 1$, $\mu(d_2) = -1$ and hence $\mu(d_1) + \mu(d_2) = 0$ and $|\mu(d_1)| + |\mu(d_2)| = 2$.

Case II. If n is some power of some prime number p, then $k = 1$. Let $n = p^a$. Then, all the divisors of n are $p, p^2, p^3, ..., p^a$. Here, $\mu(1) = 1$, $\mu(p) = -1$, $\mu(p^2) = \mu(p^3) = ... = \mu(p^a) = 0$.

Hence, $\mu(d_1) = 1$, $\mu(d_2) = -1$, $\mu(d_3) = \mu(d_4) = ... = \mu(d_m) = 0$. Therefore,

$$\mu(d_1) + \mu(d_2) + ... + \mu(d_m) = 0 \text{ and}$$

$$| \mu(d_1) | + | \mu(d_2) | + ... + | \mu(d_m) | = 2 = 2^k$$

Case III. If n is a product of k prime numbers, *i.e.*, $n = p_1p_2 ..., p_k$, then $p, p_1, p_2, ..., p_k, p_1 p_2, p_1 p_3, p_2 p_3, ..., p_{k-1}p_k, p_1 p_2 p_3, p_1 p_2 p_4 ..., p_{k-2} p_{k-1} p_k, ..., p_1 p_2 ... p_k$ are the divisors of n. Now

$$\mu(1) = 1$$

$$\mu(p_1) = \mu(p_2) = ... = \mu(p_k) = -1,$$

$$\mu(p_1 p_2) = \mu(p_1 p_3) = ... = \mu(p_{k-1} p_k) = 1 \text{ and these are } C_2^k$$

in numbers,

$$\mu(p_1 p_2 p_3) = ... = \mu(p_{k-2} p_{k-1} p_k) = 1$$

and these are C_3^k in numbers,

... ...

$$\mu(p_1p_2 ... p_r) = ... = (-1)^r$$

and these are C_r^k in number.

... ...

Thus, $\mu(d_1) + \mu(d_2) + ... + \mu(d_m)$

$$= \mu(1) + \sum_i \mu(p_i) + \sum_{i \neq j} \mu(p_i p_j) + \sum_{i \neq j \neq k} \mu(p_i p_j p_k) + ...$$

$$= 1 - k + C_2^k - C_3^k + ... = (1 - 1)^k = 0$$

Similarly,

$$| \mu(d_1) | + | \mu(d_2) | + ... + | \mu(d_m) |$$

$$= | \mu(1) | + \sum_i | \mu(p_i) | + \sum_{i \neq j} \left| \mu(p_i p_j) \right| + ...$$

$$= 1 + C_1^k + ... = (1 + 1)^k = 2^k$$

8.3. Let f be a multiplicative function, *i.e.*, If $(n_1, n_2) = 1$, then $f(n_1 n_2) = f(n_1)\, f(n_2)$. Then the function g defined by

$$g(n) = \sum_{d/n} f(d)$$

is also a multiplicative function.

Proof. Let $(n, n') = 1$, $d \mid n$, $d' \mid n'$. Then, $(d, d') = 1$. By considering each such pair (d, d') we get all the divisors, $c = dd'$, of nn'. Now, from the definition of g,

$$g(nn') = \sum_{c/nn'} f(c) = \sum_{d/n,\, d'/n'} f(dd')$$

$$= \sum_{d/n,\, d'/n'} f(d) f(d') = \sum_{d/n} f(d) \sum_{d'/n'} f(d')$$

$$= g(n)\ g(n')$$

8.4. The Mobius inversion formula:

Let $g(n) = \sum_{d/n} f(d)$, then $f(n) = \sum_{d/n} \mu(d)\, g(n/d)$

Proof. First observe that

$$\sum_{d/n} \mu(d)\, g(n/d) = \sum_{dd'=n} \mu(d)\, g(d')$$

$$= \sum_{dd'=n} \mu(d) \left(\sum_{c/d'} f(c) \right) = \sum_{cd/n} \mu(d)\, f(c)$$

$$= \sum_{c/n} f(c) \left(\sum_{d/\frac{n}{c}} \mu(d) \right)$$

But by the previous section if $\frac{n}{c} \neq 1$, then $\sum_{d \mid \frac{n}{c}} \mu(d) = 0$ and if $\frac{n}{c} = 1$,

then $\sum_{d \mid \frac{n}{c}} \mu(d) = \mu(1) = 1$

So, $$\sum_{c|n} f(c)\left(\sum_{d\left|\frac{n}{c}\right.} \mu(d)\right) = f(1).0 + \ldots + f(n) = f(n)$$

and hence,

$$\sum_{d/n} \mu(d)\, g\left(\frac{n}{d}\right) = f(n)$$

Conversely, if $f(n) = \sum_{d/n} \mu(d)\, g\left(\frac{n}{d}\right)$, then

$$g(n) = \sum_{n/d} f(d)$$

Proof. Clearly

$$\sum_{d|n} \mu(d)\, g\left(\frac{n}{d}\right) = \sum_{d|n} \mu\left(\frac{n}{d}\right) g(d)$$

So $$f\left(\frac{n}{d}\right) = \sum_{c\left|\frac{n}{d}\right.} \mu\left(\frac{n}{cd}\right) g(c)$$

$$\sum_{d|n} f(d) = \sum_{d|n} f\left(\frac{n}{d}\right) = \sum_{d|n} \sum_{c\left|\frac{n}{d}\right.} \mu\left(\frac{n}{cd}\right) g(c)$$

$$= \sum_{cd|n} \mu\left(\frac{n}{cd}\right) g(c) = \sum_{c|n} g(c)\left(\sum_{d\left|\frac{n}{c}\right.} \mu\left(\frac{n}{cd}\right)\right)$$

As before if $c \neq n$, $\sum_{d\left|\frac{n}{c}\right.} \mu\left(\frac{n}{cd}\right) = 0$ and if

$$c = n, \quad \sum_{d\left|\frac{n}{c}\right.} \mu\left(\frac{n}{cd}\right) = \mu(1) = 1$$

So as before $\sum_{d|n} f(d) = g(n)$.

8.5. Let n be any given positive integer. Let its number of divisors including 1 and n be denoted by $d(n)$. Let $\sigma(\nu)$ denote the sum of its divisors and $\sigma_k(n)$ the sum of the kth powers of its divisors. So, if $m_1, m_2, m_3, ..., m_g$ are all the divisors of n, then

$$d(n) = k = (m_1)^0 + (m_2)^0 + ... + (m_g)^0$$

$$\sigma(n) = m_1 + m_2 + ... + m_g$$

$$\sigma_k(n) = (m_1)^k + (m_2)^k + ... + (m_g)^k$$

We adopt the following procedure for the determination of these numbers.

Let $p_1, p_2, ..., p_l$ be the prime divisors of n and let $n = p_1^{a_1} p_2^{a_2} ... p_l^{a_l}$. Then, all the divisors of n including 1 and n can be represented as:

$$m = p_1^{b_1} p_2^{b_2} ... p_l^{b_l}.$$

Where $b_1 \in \{0, 1, 2, ..., a_1\}$,

$b_2 \in \{0, 1, 2, ..., a_2\}$,

... ...

$b_l \in \{0, 1, 2, ..., a_l\}$

Thus, b_1 can take $(a_1 + 1)$ possible values and b_2 can take $(a_2 + 1)$ possible values and so on. Therefore, b_1, b_2 can take $(a_1 + 1)(a_2 + 1)$ possible values. If we proceed in this way, we observe that $\{b_1, b_2, ... b_l\}$ can take $(a_1 + 1)(a_2 + 1)...(a_l + 1)$ possible distinct values and we denote this product by $\prod_{i=1}^{l}(a_i + 1)$.

Thus $$d(n) = \prod_{i=1}^{l}(a_i + 1)$$

Again, the sum of these divisors

$$\sigma(n) = \sum_{b_1=0}^{a_1} \sum_{b_2=0}^{a_2} ... \sum_{b_l=0}^{a_l} p_1^{b_1} p_2^{b_2} ... p_l^{b_l}$$

$$= (1 + p_1 + p_1^2 + ... + p_1^{a_1})(1 + p_2 + p_2^2$$

$$... + p_2^{a_2}) ... (1 + p_l + p_l^2 + ... + p_l^{a_l}).$$

$$= \prod_{i=1}^{l} \left(\frac{p_i^{ai+1} - 1}{p_i - 1} \right)$$

Similarly, $\sigma_k(n) = \prod_{i=1}^{l} (1 + p_i^k + p_i^{2k} + ... + p_i^{a_{ik}})$

$$= \prod_{i=1}^{l} \left(\frac{p_i^{(a_i+1)k} - 1}{p_i^k - 1} \right)$$

8.6. Let f be an arithmetical function and let s_0 be a fixed number. For each $s > s_0$, let the infinite series

$f(1) + \frac{f(2)}{2^s} + \frac{f(3)}{3^s} + ... + \frac{f(n)}{n^s} + ...$ be convergent.

We denote this series by $x(s)$ and write

$$x(s) = \sum_{n=1}^{\infty} \frac{f(n)}{n^s}, \qquad (s > s_0)$$

In this case x is said to be the Generating Function of f.

Further, if $f(1) = 1$ and f is a multiplicative arithmetical function, then for any prime number p, consider

$$x_p(s) = 1 + \frac{f(p)}{p^s} + \frac{f(p^2)}{p^{2s}} + ... \frac{f(p^r)}{p^{rs}} + ...$$

Let this given positive integer N be represented as the product of prime numbers as follows:

$$N = p_1^{r_1} p_2^{r_2} ... p_k^{rk}.$$

Then, $\frac{f(p_1^{r_1})}{p_1^{r_1 s}} \cdot \frac{f(p_2^{r_2})}{p_2^{r_2 s}} \cdots \frac{f(p_k^{rk})}{p_k^{r_k s}}$

$$= \frac{f(p_1^{r_1} p_2^{r_2} \cdots p_k^{rk})}{p_1^{r_1} p_2^{r_2} \cdots p_l^{rk}} = \frac{f(\mathrm{N})}{\mathrm{N}^s}$$

is a term in the infinite series which represents $x_{p_1}(s)\, x_{p_2}(s) \ldots x_{p_k}(s)$ and is obtained by multiplying the infinite series which represents each $x_p(s)$. In other words, those terms of $x(s) = \sum_1^\infty \frac{f(n)}{n^s}$ which can be represented as $n = p_1^{r_1} p_2^{r_2} \ldots p_k^{rk}$, $\frac{f(n)}{n^s}$ will be one term in the infinite series representing $x_{p_1}(s)\, x_{p_2}(s) \ldots x_{p_k}(s)$

On the other hand, if M is any positive integer and if all the prime numbers not greater than M are arranged in order of magnitude as follows:

$$2,\ 3,\ 5,\ 7,\ 11,\ \ldots\ p$$

then each term in the infinite series representing $x_2(s)\, x_3(s)\, x_5(s) \ldots x_p(s)$ is a term of the infinite series representing $x(s)$.

So, if both are convergent, then

$$x(s) = x_2(s)\ x_3(s)\ x_5(s) \ldots x_p(s) \ldots$$

We write this as

$$x(s) = \prod_{\substack{p \\ p \text{ prime}}} x_p(s)$$

8.7. For $s > 1$, consider the function ξ defined by

$$\xi(s) = \sum_{n=1}^{\infty} \frac{1}{n^s}$$

In this case, $f(n) = 1$ for all n. This is called the *Riemann Zeta function*. In this case, $X_p(s)$ is given by

$$X_p(s) = 1 + \frac{1}{p^s} + \frac{1}{p^{2s}} + \frac{1}{p^{3s}} + \ldots + \frac{1}{p^{rs}} + \ldots = \frac{1}{1 - p^{-s}}$$

Since $X(s) = p(s)$ and

$$X(s) = \coprod_p X_p(s) \text{ (}p\text{ prime) we get}$$

$$p(s) = \coprod_{p \text{ prime}} (1 - p^{-s})^{-1}.$$

Note that s is the Generating Function of $f \equiv 1$.

8.8. We now determine the generating function of μ_s for every prime p, let

$$X_p(s) = 1 - p^{-s}$$
$$= 1 + \mu(p)\, p^{-s} + \mu(p^2)\, p^{-2s} + \ldots + \mu(p^r)\, p^{-rs} + \ldots$$

Since $\mu(1) = 1$ and μ is multiplicative

$$X(s) = \sum_{n=1}^{\infty} \mu(n)\, n^{-s}$$

So,
$$\frac{1}{\rho(s)} = \coprod_p (1 - p^{-s})$$

$$= \prod_p \{1 + \mu(p)\, p^{-s} + \mu(p^2)\, p^{-2s} + \ldots$$

$$+ \mu(p^r)\, p^{-rs} + \ldots$$

$$= \coprod_p X_p(s) = X(s) = \sum_{n=1}^{\infty} \mu(n) n^{-s}$$

$$= 1 + \mu(2)2^{-s} + \mu(3)3^{-s} + \ldots + \mu(n)n^{-s} + \ldots$$

Hence, the generating function of μ is $\dfrac{1}{\rho}$.

8.9. In this section, we determine the generating function of the *Euler's function* ϕ. We first show that if F and G are respectively the generating functions of f and g and if

$$h(n) = \sum_{lm=n} f(l)\, g(m),$$

then FG is the generating function of h.

Let $$n = p_1^{r_1} p_2^{r_2} \ldots p_k^{rk} = \prod_{r=1}^{k} \left(p_r^{a_r}\right)$$

Then $$\phi(n) = n. \sum_{r=1}^{k} \left(1 - \frac{1}{p^r}\right)$$

$$= n. \sum_{r=1}^{k} \left\{ \frac{\mu(p_r^{0})}{p_r^{n}} + \frac{\mu(p_r)}{p_r} + \frac{\mu(p_r^{2})}{p_r^{2}} + \ldots \frac{\mu(p_r^{a_r})}{p_r^{a_r}} \right\}$$

$$= n \left(\sum_{0 \le e_1 \le a_1} \ldots \sum_{0 \le e_r \le a_r} \ldots \sum_{0 \le e_k \le a_k} \frac{\mu(p_1^{e_1}) \ldots \mu(p_r^{e_r}) \ldots \mu(p_k^{e_k})}{p_1^{e_1} \ldots p_r^{e_r} \ldots p_k^{e_k}} \right)$$

$$= n \left(\sum_{e_1 \ldots e_k} \frac{\mu(p_1^{e_1} p_2^{e_2} \ldots p_k^{e_k})}{p_1^{e_1} p_2^{e_2} \ldots p_k^{e_k}} \right)$$

$$= n \sum_{d/n} \frac{\mu(d)}{d} = \sum_{d/n} (n/d)\, \mu(d) = \sum_{dd'=n} d'\mu(d) = \sum_{lm=n} l.\mu(m)$$

Again $$\sum_{l=1}^{\infty} \frac{1}{l^3} = \sum_{l=1}^{\infty} \frac{1}{l^{e-1}} = \rho^{(s-1)}$$

and $$\sum_{m=1}^{\infty} \frac{\mu(m)}{m^2} = \frac{1}{\rho(s)}$$

Take $f(l) = l$, $g(m) = \mu(m)$.

Since
$$F(s) = f(1) + \frac{f(2)}{2^s} + \ldots + \frac{f(l)}{l^s} + \ldots$$

$$G(s) = g(1) + \frac{g(2)}{2^s} + \ldots + \frac{g(m)}{m^s} + \ldots, \text{ we have}$$

$$F(s)\, G(s) = \left\{\sum_{l=1}^{\infty} f(l).\, l^{-s}\right\}\left\{\sum_{m=1}^{\infty} g(m)\, m^{-s}\right\}$$

$$= \sum_{l,m} f(l)\, g(m)\, l^{-s}\, m^{-s}$$

$$= \sum_{l,m} f(l)\, g(m)\, (lm)^{-s}$$

$$= \sum_{n} n^{-s}\left\{\sum_{lm=n} f(l)\, g(m)\right\} = \sum_{n=1}^{\infty} h(n)\, n^{-s}$$

Hence, the generating function of h is FG.

Now, let $h(n) = \varphi(n)$; and let the prime factors of n be given by p_1, p_2, ..., p_k. Then,

$$\varphi(n) = n\left(1 - \frac{1}{p_1}\right)\left(1 - \frac{1}{p_2}\right)\ldots\left(1 - \frac{1}{p_k}\right)$$

$$h(n) = \sum_{lm=n} f(l)\, g(m) = \sum_{lm=n} l.\, \mu(m) = \varphi(n)$$

But the generating functions of $f(l)$ and $g(m)$ are respectively given by $\rho(s-1)$ and $\frac{1}{\rho(s)}$. Therefore, the generating function of $h(n) = \varphi(n)$ is given by

$$\left[\rho(s-1) . \frac{1}{\rho(s)}\right]$$

i.e., $$\frac{\rho(s-1)}{\rho(s)} = \sum_{n=1}^{\infty} \frac{\varphi(n)}{n^s}.$$

8.10. By using the above procedure we shall show that if $s > k + 1$, then

$$\rho(s)\,\rho(s-k) = \sum_{n=1}^{\infty} \frac{\sigma k(n)}{n^2}$$

Let $f(l) = 1$ and $g(m) = m^k$.

Then, $$h(n) = \sum_{lm=n} f(l)\,g(m) = \sum_{lm=n} 1.m^k = \sum_{m|n} m^k = \sigma_k(n)$$

But the generating function of $f(l)$

$$= \sum \frac{f(l)}{l^s} = \sum \frac{l}{l^s} = \rho(s)$$

and the generating function of $g(m)$

$$= \sum \frac{g(m)}{m^s} = \sum \frac{m^k}{m^s} = \sum \frac{1}{m^{s-k}} = s(s-k)$$

So the generating function of $\sigma k(n) = h(n)$ is $\rho(s)\,\rho(s-k)$ and this completes the proof.

Corollary 1. If $k = 0$, then

$$\rho^2(s) = \sum_{n=1}^{\infty} \frac{d(n)}{n^s} \quad (s > 1)$$

Corollary 2. If $k = 1$, then

$$\rho(s)\rho(s-1) = \sum_{n=1}^{\infty} \frac{\sigma(n)}{n^s} \; (s > 2)$$

CHAPTER 9

Roots of Polynomials

In this chapter, we shall be concerned with the polynomials over the complex field C and in particular, over the real field R, rational field Q and ring of integers Z. We shall discuss some important properties of roots of such polynomials and relation between the roots and the coefficients.

There are several methods or rules given by several eminent mathematicians such as Gauss, Newton, Euler, Fourier, Ferrai, Budan, Strum, Horner for determination of roots of general quadratic, cubic and biquadratic equations.

We shall not discuss any of these methods in this book. The interested reader may refer to any good text on Theory of Equations. It should be noted that the solutions of polynomial equations of degrees higher than 4 are given only in some special cases. For example, the solutions of the equation $x^{17}-1 = 0$ was discovered by the famous mathematician Gauss. The reason is that any polynomial of degree greater than or equal to 5 is not solvable by radicals. We cannot prove this result in this introductory book because it needs the ideas of Galois theory which we do not intend to discuss here. The reader is referred to any advanced text on abstract algebra which deals with Galois theory.

Another important result which is called the fundamental theorem of algebra states that the complex number field C is algebraically closed, in other words, every polynomial of degree n over C (or in particular, over R or Q or Z) has exactly n roots (and hence all the roots) in C. The proof of this result uses techniques of complex analysis and we do not intend to prove this result here.

9.1. Properties of Roots

The following theorems (Theorem 60 – 64) describe some basic properties of roots of polynomials and the relation between roots and coefficients.

Theorem 60

Let $f(x) = a_0 + a_1x + ... + a_nx^n$ be a polynomial over Z. If $\frac{p}{q} \in Q$ is a root of $f(x)$, then $p \mid a_0$ and $q \mid a_n$.

Proof. If $\frac{p}{q}$ is a root of $f(x)$, then

$$a_0 + a_1\left(\frac{p}{q}\right) + ... + a_n\left(\frac{p}{q}\right)^n = 0$$

or $\quad a_0q^n + a_1pq^{n-1} + ... + a_np^n = 0$

or

$$\begin{aligned} a_0q^n &= - a_1pq^{n-1} - ... - a_np^n \\ &= p(-a_1q^{n-1} - ... - a_{n-1}p^{n-2}q - a_np^{n-1}) \end{aligned}$$

Thus, $p \mid a_0q^n$ and since p and q are relatively prime $p \mid a_0$. Also, we have

$$\begin{aligned} a_np^n &= -a_0q^n - a_1pq^{n-1} - ... - a_{n-1}p^{n-1}q \\ &= q(-a_0q^{n-1} - a_1pq^{n-2} - ... - a_{n-1}p^{n-1}) \end{aligned}$$

Thus, $q \mid a_np^n$ and since p and q are relatively prime $q \mid a_n$ and this completes the proof.

Theorem 61

If a polynomial $f(x)$ with real coefficients has $\alpha + i\beta$ as one of the roots, then $\alpha - i\beta$ is also a root, in other words, if complex roots occur, then they occur in pair.

Proof. Let $f(x) \in R[x]$, the set of all polynomials with real coefficients, be of degree n and let $\alpha + i\beta$ be a root of $f(x)$, *i.e.*, $f(\alpha + i\beta) = 0$. Observe that

$$\{x - (\alpha + i\beta)\}\{x - (\alpha - i\beta)\} = (x - \alpha)^2 + \beta^2$$

Now, divide $f(x)$ by $\{(x - \alpha)^2 + \beta^2\}$.

By division algorithm, we obtain

$$f(x) = \{(x - \alpha)^2 + \beta^2\}\ \phi(x) + (k_1x + k_2)$$

where $\phi(x)$ is a polynomial over R of degree $(n - 2)$ and k_1, k_2 are two real constants. Now, $\alpha + i\beta$ is a root of $f(x)$, *i.e.*, $f(\alpha + i\beta) = 0$.

Therefore

$$\begin{aligned} 0 &= \{(\alpha + i\beta - \alpha)^2 + \beta^2\}\ \phi(\alpha + i\beta) + k_1(\alpha + i\beta) + k_2 \\ &= (-\beta^2 + \beta^2)\ \phi(\alpha + i\beta) + k_1(\alpha + i\beta) + k_2 \\ &= k_1(\alpha + i\beta) + k_2 \\ &= (k_1\alpha + k_2) + i\beta k_1 \end{aligned}$$

Consequently, $k_1\alpha + k_2 = 0$

and $\quad k_1\beta = 0$, where $\beta \neq 0$

Therefore, $k_1 = 0$ and so $k_2 = 0$

Hence, $\quad f(x) = \{(x - \alpha)^2 + \beta^2\}\phi(x)$

$$= \{x - (\alpha + i\beta)\}\ \{x - (\alpha - i\beta)\}\phi(x)$$

Now, since $\{x - (\alpha - i\beta)\}$ is also a factor of $f(x)$, this shows that $(\alpha - i\beta)$ is also a root of $f(x)$. This completes the proof.

Theorem 62

Let $f(x)$ be a polynomial with rational coefficients. Then, its irrational roots occur in pair.

Proof. Let $f(x)$ be a polynomial of degree n over the field Q of rational numbers. Let $a + \sqrt{b}$ be an irrational root of $f(x)$.

Observe that

$$(x - a - \sqrt{b})\,(x - a + \sqrt{b}) = (x - a)^2 - b$$

Divide $f(x)$ by $(x - a)^2 - b$. Let $f(x) = \{(x - a)^2 - b\}\phi(x) + k_1x + k_2$ where k_1 and k_2 are rationals. Since $a + \sqrt{b}$ is a root of $f(x)$, we have

$$\begin{aligned} 0 = f(a + \sqrt{b}) &= \{(a + \sqrt{b} - a)^2 - b\}\ \phi(a + \sqrt{b}) \\ &\qquad + k_1(a + \sqrt{b}) + k_2 \\ &= k_1(a + \sqrt{b}) + k_2 \\ &= (k_1 a + k_2) + k_1\sqrt{b} \end{aligned}$$

Therefore $k_1a + k_2 = 0$ and $k_1 = 0$

Consequently, $k_2 = 0$

Thus, $f(x) = \{(x - a)^2 - b\}\phi(x)$

$$= [\{x-(a+\sqrt{b})\}\{x-(a-\sqrt{b})\}]\,\phi(x)$$

Therefore, $a - \sqrt{b}$ is also a root of $f(x)$ and this completes the proof.

In the following theorem, we shall discuss some properties of the polynomial $x^n - 1$.

Theorem 63

(*i*) If α is a root of $x^n - 1$, then α^r is also a root of $x^n - 1$, where r is a positive integer.

(*ii*) If m and n are relatively prime then the polynomials $x^n - 1$ and $x^m - 1$ do not have any common root except 1.

(*iii*) If α is an imaginary root of $x^n - 1$ where n is prime, then all its roots are $1, \alpha, \alpha^2, \ldots, \alpha^{n-1}$.

Proof

(*i*) Since α is a root of $x^n - 1$, we have

$$\alpha^n = 1.$$

Therefore, $(\alpha^r)^n = (\alpha^n)^r = 1^r = 1$

and hence α^r is a root of $x^n - 1$.

(*ii*) Since m and n are relatively prime, we can choose integers a and b such that $mb - na = \pm 1$. Assume that α is a root of both the polynomials $x^n - 1$ and $x^m - 1$.

Then, $\alpha^m = 1$ and $\alpha^n = 1$

By (*i*) we have

$$\alpha^{mb} = 1 \text{ and } \alpha^{na} = 1.$$

Therefore, $\alpha^{mb-na} = 1$, *i.e.*, $\alpha = \pm 1$, *i.e.*, $\alpha = 1$ and this completes the proof.

(*iii*) Let α be a root of $x^n - 1$. By (*i*) α^r is also a root of $x^n - 1$ where r is a positive integer. Since $x^n - 1$ is a polynomial of degree n, it has exactly n roots in the field of complex numbers. Again, 1, α, α^2, ... are the roots of $x^n - 1$. In order to prove the theorem, it is enough to show that 1, α, α^2, ... α^{n-1} are all distinct. If possible assume that $\alpha^r = \alpha^m$ with $r < m < n$. Then $\alpha^{m-r} = 1$. Hence α is a root of $x^{m-r} - 1$. In other words α is a common root of $x^n - 1$ and $x^{m-r} - 1$. But m and r are both less than n and n is a prime, $(m - r)$ and n are relatively prime. Therefore there is no root common to $x^n - 1$ and $x^{m-r} - 1$ except 1. Hence, $\alpha^r \neq x^m$ for $r \neq m$ and this completes the proof of the theorem.

The following theorem gives the relations between the roots and coefficients of a polynomial.

Theorem 64

Let $f(x) = a_0 + a_1x + a_2x^2 + ... + a_nx^n$ be a polynomial over the field Q or R or C. Let $\alpha_1, \alpha_2, ..., \alpha_n$ be all the distinct roots of $f(x)$ all lying in C. Let

$$S_1 = \alpha_1 + \alpha_2 + ... + \alpha_n = \Sigma\alpha_1$$
$$S_2 = \alpha_2\alpha_2 + ... = \Sigma\alpha_1\alpha_2$$
$$S_3 = \Sigma\alpha_1\alpha_2\alpha_3$$
$$........$$
$$........$$
$$S_n = \alpha_1\alpha_2...\alpha_n$$

Then,

$$S_j = (-1)^j \frac{a_{n-j}}{a_n}, \ j = 1, 2, ..., n$$

Proof. Since $\alpha_1, \alpha_2, ..., \alpha_n$ are the roots of $f(x)$, we have

$$f(x) = a^n (x - \alpha_1)(x - \alpha_2) .... (x - \alpha_n)$$

or $a_0 + a_1x + ... + a_nx^n$

$$= a_n(x - \alpha_1)(x - \alpha_2)...(x - \alpha_n)$$
$$= a_n[x^n - (\Sigma\alpha_1)x^{n-1} + (\Sigma\alpha_1\alpha_2)x^{n-2} + ... + (-1)^n\alpha_1\alpha_2...\alpha_n]$$
$$= [\alpha^n - S_1x^{n-1} + (-1)^2S_2x^{n-2} + ... + (-1)^nS_n]$$

Now, comparing the coefficients of various powers of x on both sides, we get

$$a_{n-1} = -a_n S_1 \qquad \text{or} \qquad S_1 = -\frac{a_{n-1}}{a_n}$$

$$a_{n-2} = (-1)^2 a_n S_2 \quad \text{or} \quad S_2 = (-1)^2 \frac{a_{n-2}}{a_n}$$

$$a_{n-j} = (-1)^j a_n S_j \qquad \text{or} \quad S_j = (-1)^j \frac{a_{n-j}}{a_n}$$

9.2. Solution of Cubic Equation

We shall discuss below the solutions of general cubic equations by radical method. The general cubic equation is given by

$$x^3 + px^2 + qx + r = 0$$

This can be transformed into a simpler equation in the form

$$x^3 + px + r = 0$$

In order to solve this equation, put $x = y + z$. Then, we have

$$\begin{aligned} x^3 &= y^3 + z^3 + 3yz(y + z) \\ &= y^3 + z^3 + 3yzx \end{aligned}$$

and the equation takes the form

$$y^3 + z^3 + (3yz + q)x + r = 0$$

At present, y and z are any two quantities subjected to the condition that their sum is equal to one of the roots of the given equation. If we further assume that they satisfy the equation $3yz + q = 0$, then they are completely determined. We thus obtain

$$y^3 + z^3 = -r, \; y^3z^3 = -\frac{q^3}{27}$$

Hence, y^3 and z^3 are the roots of the quadratic equation.

$$t^2 + rt - \frac{q^3}{27} = 0$$

Solving this equation, we get

$$y^3 = -\frac{r}{2} + \sqrt{\frac{r^2}{4} + \frac{q^3}{27}}$$

$$z^3 = -\frac{r}{2} - \sqrt{\frac{r^2}{4} + \frac{q^3}{27}}$$

Thus, we get the value of x from the relation $x = y + z$ as follows:

$$x = \left\{\frac{-r}{2} + \sqrt{\frac{r^2}{4} + \frac{q^3}{27}}\right\}^{1/3} + \left\{\frac{-r}{2} - \sqrt{\frac{-r}{4} + \frac{q^3}{27}}\right\}$$

9.3. Solution of Biquadratic Equation

We shall next discuss the solution of biquadratic equation by radical method. This was first obtained by Ferrari.

Let the equation be

$$x^4 + 2px^3 + qx^2 + 2rx + s = 0$$

Add $(ax + b)^2$ to both the sides, the numbers a and b to be determined later so that the left-hand side becomes a perfect square.

Then, we have

$$x^4 + 2px^2 + (q + a)^2x^2 + 2(r + ab)x + s + b^2 = (ax + b)^2$$

Suppose that the left-hand side is equal to $(x^2 + px + k)^2$. Then by comparing the coefficients on both the sides, we obtain

$$p^2 + 2k = q + a^2,\ pk = r + ab,\ k^2 = s + b^2$$

By eliminating a and b from the equations, we obtain

$$(pk - r)^2 = (2k + p^2 - q)\ (k^2 - s)$$

or $\quad 2k^2 - qk^2 + 2(pr - s)k - p^2s + qs - r^2$

For this cubic equation, one real value of k can always be found, thus a and b are known.

Also, $$(x^2 + px + k)^2 = (ax + b)^2$$

or $$x^2 + px + k = \pm(ax + b)$$

Now, the values of x are to be obtained from the two quadratic equations

$$x^2 + (p - a)x + (k - b) = 0$$

and $$x^2 + (p + a)x + (k + b) = 0$$

and this completes the solution.

Example

(*a*) Determine a polynomial whose roots are –3, –1, 4, 5.

Since –3, –1, 4 and 5 are the roots, the polynomial $f(x)$ is given by

$$\begin{aligned} f(x) &= (x + 3)(x + 1)(x - 4)(x - 5) \\ &= x^4 + 5x^3 - 13x^2 + 53x + 60 \end{aligned}$$

(*b*) If one root of $x^4 + 2x^3 - 5x^2 + 6x + 2$ is $-2 + \sqrt{3}$, find the other roots.

Since irrational roots occur in pair and since $-2 + \sqrt{3}$ is a root, $-2 - \sqrt{3}$ is another root. Therefore,

$(x + 2 + \sqrt{3})(x + 2 - \sqrt{3}) = x^2 + 4x + 1$ is a factor of the given polynomial. Dividing the given polynomial by $x^2 + 4x + 1$, we get $x^2 - 2x + 2$. Hence, the other roots are the roots of $x^2 - 2x + 2$, given by

$$\frac{2 \pm \sqrt{4 - 8}}{2} = 1 \pm i$$

Observe that the two complex roots also occur in pair.

(*c*) If the sum of two roots of $x^3 - 5x^2 - 16x + 80$ is zero, then find all its roots.

This is a cubic equation with integer coefficients and hence it has exactly three roots in the complex field. Let the roots be a, b and c.

Then,
$$a + b + c = 5 \qquad \text{... (1)}$$
$$ab + bc + ca = -16 \qquad \text{... (2)}$$
$$abc = -80 \qquad \text{... (3)}$$

According to the given condition, we may write
$$b + c = 0 \qquad \text{... (4)}$$

It now follows from (1) and (4) that $a = 5$. Then, from (3), we get $bc = -16$. Therefore, $b - c = \sqrt{(b+c)^2 - 4bc} = \pm 8$, hence $b = 4$ and $c = -4$. So, all the roots are 5, 4, –4.

(*d*) If the roots of the polynomial $27x^3 + 42x^2 - 28x - 8$ are in GP, then find all the roots.

Since the roots are in GP, we can assume without any loss of generally that they are $\frac{a}{b}$, a and ab.

Therefore,
$$\frac{a}{b} + a + ab = \frac{42}{27} \qquad \text{... (1)}$$
$$\frac{a^2}{b} + a^2 + a^2b = -\frac{28}{27} \qquad \text{... (2)}$$
$$\frac{a}{b} \,.\, a \,.\, ab = \frac{8}{27} \qquad \text{... (3)}$$
$$a = \frac{2}{3}$$

$\therefore$ On putting this value in (1), we get
$$\frac{1}{b} + 1 + b = -\frac{21}{9}$$

or
$$b^2 + \frac{10}{3}b + 1 = 0$$

or $$b = -3 \quad \text{or} \quad -\frac{1}{3}$$

Hence, the roots are $-2, \frac{2}{3}, -\frac{2}{9}$.

(*e*) If a, b and c are the roots of $x^3 + px^2 + qx + r$, then find the value of Σa^2b.

Now, $$a + b + c = -p$$
$$ab + bc + ca = q$$
$$abc = -r$$

Since $(a + b + c)(ab + bc + ca) = \Sigma a^2b + 3abc$, we have

$$\Sigma a^2b = \Sigma a \, \Sigma ab - 3abc$$
$$= -p.q - 3.(-r)$$
$$= -pq + 3r$$

(*f*) If $\alpha_1, \alpha_2, ..., \alpha_n$ are the roots of $x^n + p_1x^{n-1} + ... + p_n$, then find the values of $\Sigma\frac{1}{\alpha_1}$ and $\Sigma\frac{1}{\alpha_2\alpha_2}$

We know that

$$\alpha_3\alpha_4 \ldots \alpha_n + \alpha_1\alpha_4 \ldots \alpha_n + \ldots = (-1)^{n-2}p_{n-2} \quad \ldots (1)$$
$$\alpha_2\alpha_3 \ldots \alpha_n + \alpha_1\alpha_3 \ldots \alpha_n + \ldots = (-1)^{n-1}p_{n-1} \quad \ldots (2)$$
$$\alpha_1\alpha_2\alpha_3 \ldots \alpha_n = (-1)^n p_n \quad \ldots (3)$$

By dividing (2) by (3) we get

$$\Sigma\frac{1}{\alpha_1} = \frac{1}{\alpha_1} + \frac{1}{\alpha_2} + \ldots + \frac{1}{\alpha_n} = \frac{p_{n-1}}{p_n}$$

(*g*) If a, b and c are the roots of $x^3 + px^2 + qx + r$, then find the value of $\Sigma\frac{b^2 + c^2}{bc}$.

Since a, b and c are the roots of $x^3 + px^2 + qx + r$, we have

$$\Sigma a = -p$$

$$\Sigma ab = q$$

$$abc = -r$$

Now $$\Sigma\frac{b^2+c^2}{bc} = \frac{\Sigma a^2 b}{abc}$$

Again, $$\Sigma a\, \Sigma ab = \Sigma a^2 b + 3abc$$

Therefore. $$\Sigma\frac{b^2+c^2}{bc} = \frac{\Sigma a \Sigma ab - 3abc}{abc}$$

$$= \frac{-p.q+3r}{-r}$$

$$= \frac{pq}{r} - 3$$

(*h*) If the roots of the polynomial $x^n + p_1 x^{n-1} + \ldots + p_n$ are in AP, then find them.

Let the roots of the given polynomial be

$$a,\ a + d,\ a + 2d,\ \ldots,\ a + (n - 1)d.$$

Then, $$a + (a + d) + \ldots + \{a + (n - 1)d\}$$

$$= na + \frac{n(n-1)}{2}d$$

Therefore,

$$na + \frac{n(n-1)}{2}d = -p$$

Now, $$a^2 + (a + d)^2 + \ldots + \{a + (n - 1)d\}^2$$

$$= [a + (a + d) + \ldots + \{a + (n - 1)d\}]^2 - 2\Sigma a\,(a + d)$$

$$= (-p_1)^2 - 2p_2 = p_1^2 - 2p_2$$

Now, multiply (2) by n and from this, subtract the square of (1). This gives

$$d^2\left\{\frac{n^2(n-1)(2n-1)}{6}-\frac{n^2(n-1)^2}{4}\right\} = (p_1^2 - 2p_2)n - p_1^2$$

or
$$d^2 = \frac{12\{(n-1)p_1^2 - 2np_2\}}{n^2(n^2-1)}$$

This gives d. Putting this value of d in (1), we get a, and consequently, we get all the roots.

(*i*) If the sum of two roots of $x^3 + px^2 + qx + r$ is equal to third one, prove that $p^3 - 4pq + 8r = 0$.

We have
$$\alpha_1 + \alpha_2 = \alpha_3$$

$$\alpha_1 + \alpha_2 + \alpha_3 = -p \quad (1)$$

$$\alpha_1\alpha_2 + \alpha_2\alpha_3 + \alpha_3\alpha_1 = q \quad (2)$$

$$\alpha_1\alpha_2\alpha_3 = -r \quad (3)$$

Since $\alpha_1 + \alpha_2 = \alpha_3$ from (1), we get

$$\alpha_3 = -\frac{p}{2}$$

From (2), it follows that

$$\alpha_1\alpha_2 + (\alpha_1 + \alpha_2)\alpha_3 = q$$

or
$$\alpha_1\alpha_2 + \alpha_3^2 = q$$

or
$$\alpha_1\alpha_2 + \frac{p^2}{4} = q$$

or
$$\alpha_1\alpha_2 = q - \frac{p^2}{4}$$

From (3), we get

$$\left(q - \frac{p^2}{4}\right)\left(-\frac{p}{2}\right) = -r$$

or $$(4q - p^2)(-p) = -8r$$

or $$p^3 - 4pq + 8r = 0$$

(*j*) If the sum of the roots of $2x^3 - 3x^2 + kx - 1$ is 1, then find the value of k and also find all the roots of the polynomial.

Let α_1, α_2 and α_3 be the roots of $2x^3 - 3x^2 + kx - 1$.

Now, $$\alpha_1 + \alpha_2 + \alpha_3 = \frac{3}{2} \qquad ...(1)$$

$$\alpha_1\alpha_2 + \alpha_2\alpha_3 + \alpha_3\alpha_1 = \frac{k}{2} \qquad ...(2)$$

$$\alpha_1\alpha_2\alpha_3 = \frac{1}{2} \qquad ...(3)$$

$$\alpha_1 + \alpha_2 = 1 \qquad ...(4)$$

From (1) and (4), we get

$$\alpha_3 = \frac{3}{2} - 1 = \frac{1}{2}$$

From (3), $$\alpha_1\alpha_2 = 1 \qquad ...(5)$$

From (4) and (5), we get

$$(\alpha_1 - \alpha_2)^2 = (\alpha_1 + \alpha_2)^2 - 4\alpha_1\alpha_2 = 1 - 4 = -3$$

Thus, $$\alpha_1 - \alpha_3 = \pm\sqrt{3i}$$

Therefore, α_1, α_2 and α_3 are respectively

$$\frac{1}{2}, \frac{1}{2}(1 + \sqrt{3i}) \text{ and } \frac{1}{2}(1 - \sqrt{3i}).$$

and $$\alpha_1\alpha_2 + \alpha_3(\alpha_1 + \alpha_2) = \frac{k}{2}$$

or $$1 + \frac{1}{2} = \frac{k}{2} \quad \text{or} \quad k = 3$$

(*k*) Find the roots of $2x^4 - 11x^3 + 17x^2 - 11x + 15$ in the rational field and in the complex field.

If $(p, q) = 1$ and q is positive and if $p/q \in Q$ is a root of the given polynomial, then $p \mid 15$ and $q \mid 2$. Hence,

$$p = \pm 1, \pm 3, \pm 5, \pm 15, \; q = 1, 2$$

So, the possible roots are

$$\pm 1, \pm \frac{1}{2}, \pm 3, \pm \frac{3}{2}, \pm 5, \pm \frac{5}{2}, \pm 15, \pm \frac{15}{2}$$

But among these, by verification one can observe that only $f(3) = 0$ and $f(5/2) = 0$. Hence, all the rational roots are 3 and 5/2.

Now, $$f(x) = (x - 3)(2x - 5)(x^2 + 1)$$

Hence, the roots in the complex field which are not rationals are $\pm i$. Observe that this is a polynomial over the commutative ring Z, it has two roots in Q and all the four roots in C. Also, observe that complex roots occur in pairs.

(*l*) Find the rational roots $6x^4 - 7x^3 + 6x^2 - 1$. If $\frac{p}{q} \in Q$ is a root of the given polynomial then $p \mid -1$ and $q \mid 6$.

Therefore, $$p = \pm 1 \text{ and } q = 1, 2, 3, 6$$

Hence, the possible rational roots are

$$\pm 1, \pm \frac{1}{2}, \pm \frac{1}{3}, \pm \frac{1}{6}$$

But it can be checked that only $\frac{1}{2}$ and $\frac{1}{3}$ are the rational roots of the given polynomial and

$$f(x) = 6\left(x - \frac{1}{2}\right)\left(x - \frac{1}{3}\right)(x^2 - x - 1)$$

The other two roots are real but irrational and these are $\frac{1 \pm \sqrt{2}}{2}$.

(m) Find the roots of the cubic equation $x^3 - 15x - 126 = 0$.

Let $$x = p^{1/3} + q^{1/3}$$

Then, $$x^3 = p + q + 3\,p^{1/3}q^{1/3}(p^{1/3} + q^{1/3})$$

or $$x^3 - 3p^{1/3}\,q^{1/3}x - (p + q) = 0$$

By comparing this with the given equation, we get

$$p + q = 126$$

$$p^{1/3}\,q^{1/3} = 5$$

or $$pq = 125$$

Therefore, p and q are the roots of

$$t^2 - 126t + 125 = 0$$

Now, $$\begin{aligned} 0 &= t^2 - 126t + 125 \\ &= t^2 - t - 125t + 125 \\ &= (t - 1)\,(t - 125) \end{aligned}$$

Hence, $$p = 125,\ q = 1,$$

so, $$p^{1/3} = 5,\ 5w,\ 5w^2$$

$$q^{1/3} = 1,\ w^2,\ w \qquad (\text{since } p^{1/3}\,q^{1/3} = 5)$$

Therefore, the roots of the given equation are

$$5 + 1,\ 5w + w^2,\ 5w^2 + w$$

or $$6, -3 + i2\sqrt{3}, -3 - i2\sqrt{3}$$

(n) Find the roots of $x^3 - 3x + 1 = 0$.

If $x = p^{1/3} + q^{1/3}$, then $x^3 - 3p^{1/3}\,q^{1/3}x - (p + q) = 0$.

By comparing the above equation with the given equation we get

$$p + q = -1$$

$$pq = 1$$

Therefore p and q are the roots of the quadratic equation $t^2 + t + 1 = 0$.

Therefore, $$t = \frac{-1 + i\sqrt{3}}{2} = \cos\frac{2\pi}{3} \pm i\sin\frac{2\pi}{3}$$

Hence,
$$p^{1/3} = \cos\frac{\frac{2\pi}{3}+2\pi r}{3} + i\sin\frac{\frac{2\pi}{3}+2\pi r}{3}$$

$$q^{1/3} = \cos\frac{\frac{2\pi}{3}+2\pi r}{3} - i\sin\frac{\frac{2\pi}{3}+2\pi r}{3}$$

Therefore, the roots of the given equation are

$$2\cos\frac{\frac{2\pi}{3}+2\pi r}{3}, \quad r = 0, 1, 2$$

or
$$2\cos\frac{2\pi}{9}, 2\cos\frac{8\pi}{9}, 2\cos\frac{14\pi}{9}$$

EXERCISE

1. Determine the polynomials whose roots are given as follows:

   (*i*) $1+\sqrt{2}, 3, -1$ (*ii*) $1, -2, i$

   (*iii*) $1 + 5i, 5 - i$ (*iv*) $-3, -1, \frac{5}{3}$

2. Find the GCD of the following polynomial over the field of rational numbers

   (*i*) $x^3 - 6x^2 + x + 4, x^5 - 6x + 1$

   (*ii*) $x^2 + 1, x^3 + x^2 + x + 1$

   (*iii*) $x^4 + x^3 + 2x^2 + x + 1, x^3 - 1$

3. If the sum of zeros of $x^3 - 5x^2 - 16x + 80$ is zero, then determine all its zeros.

4. If the zeros of the polynomial $x^3 - px^2 + qx - r$ are in H.P., then prove that $27r^2 - 9qr + 2q^3 = 0$.

5. If the zeros of the polynomial $x^3 + 3px^2 + 3qx + r$ are in G.P., then prove that $p^3r - q^3 = 0$.

6. In a field F if the zeros of the polynomial $x^3 + qx + r$ are α, β, γ, then prove that the zeros of the polynomial $x^3 + 6qx^2 + 9q^2x + 4q^3 + 27r^3$ are $(\alpha - \beta)^2$, $(\beta - \gamma)^2$, $(\gamma - \alpha)^2$.

7. Divide $3x^4 - 5x^3 + 10x^2 + 11x - 61$ by $(x - 3)$.

8. By synthetic division, divide $x^5 - 4x^4 + 7x^3 + 11x - 13$ by $(x - 5)$

9. If one root of $x^3 + 4x^2 + x - 6 = 0$ is -2, by synthetic division find the other two roots.

10. If one root of $x^3 - 2x^2 - 2x - 3 = 0$ is 3, then by synthetic division, find the other two roots.

11. Prove that

(*i*) $x^2 + x + 1$ is irreducible over Z_2.

(*ii*) $x^2 + 1$ is irreducible over Z_7.

(*iii*) $x^3 - 9$ is irreducible over Z_{31}.

(*iv*) $8x^3 - 6x - 1$ is irreducible over Q.

(*v*) $x^3 - 2$ is irreducible over Q.

(*vi*) $x^3 + x^2 - 2x - 1$ is irreducible over Q.

(*vii*) $x^4 + 2x + 2$ is irreducible over Q.

12. If a is rational and $(x - a)$ divides an integer monic polynomial, then prove that a must be an integer.

13. If $f(x) \in F[x]$, $F = Z_p$, p prime and $f(x)$ is irreducible over F of degree n, prove that $F(x)/f(x)$ is a field with p^n elements.

14. If p is prime, prove that the polynomial $x^n - p$ is irreducible over Q.

15. Prove that the polynomial $1 + x + ... + x^{p-1}$, p prime, is irreducible over Q.

[*Hint*: Consider the polynomial $1 + (x + 1) + ... + (x + 1)^{p-1}$ and use the Eisenstein criterion].

CHAPTER 10

Valued Fields

Let F be any field. A mapping $|.| : F \to R^+$ is said to be a valuation on F if it has the following properties: for $a, b \in F$,

(*i*) $|a| = 0$ iff $a = 0$

(*ii*) $|ab| = |a|\ |b|$

(*iii*) $|a + b| \leq |a| + |b|$.

A field F along with a valuation defined on it is called a 'field with a valuation' or a 'valued field' or 'a valuated field'.

Remark: It follows from (*ii*) that $|.|$ is a homomorphism of $F^* = F - \{0\}$ the multiplicative group of positive real numbers. Therefore, if $a \in F$ with $a^n = 1$ for some $n > 0$, then $|a| = 1$. In particularm we have $|-1| = |1| = 1$

A valuation on F is said to be trivial, if for all $a \in F^*$, $|a| = 1$.

Examples:

(*i*) Let F = Q or R or C. Let $|a|$ = usual absolute value of a.

(*ii*) Let F = Q and let p be a given prime number.

For $0 \neq r = \dfrac{a}{b} = \in Q$, let $|r|_p = p^{v-u}$, where p^u and p^v are the exact powers of p dividing a and b respectively.

10.1. Definition: A valuation on a field F is said to be an Archimedean valuation if there exists a positive integer n such that $|n.1| > 1$.

Otherwise the valuation is said to be nonarchimedean. For a nonarchimedean valuation we have $|n.1| \leq 1$for every positive integer n.

Theorem 65: If F is a nonarchimedean-valued field with a nonarchimedean valuation defined on it, then

$$|a + b| \leq \max (|a|, |b|)$$

Proof. For any positive integer n and for $a, b \in$ F we have by Binomial Theorem that

$$(a + b)^n = a^n + \binom{n}{1} a^{n-1} b + \ldots + b^n$$

Therefore
$$|a + b|^n \leq |a|^n + \binom{n}{1} |a|^{n-1} |b| + \ldots + |b|^n$$

$$\leq |a|^n + |a|^{n-1} |b| + \ldots + |b|^n$$

$$\leq (n + 1) \max (|a|^n, |b|^n)$$

Taking the nth root on both sides, we have

$$|a + b| \leq (n + 1)^{1/n} \max (|a|, |b|)$$

and this completes the proof.

Remarks:

(*i*) The valuation given in Example (*i*) is archimedean, whereas that in Example (*ii*) is nonarchimedean.

(*ii*) It follows from Theorem (10.2) that a mapping. $|| : F \to R^+$ is nonarchimedean valuation if for $a, b, \in$ F,

(*i*) $|a| = 0$ iff $a = 0$

(*ii*) $|ab| = |a| \, |b|$

(*iii*) $|a + b| \leq \max (|a|, |b|)$

(*iii*) The inequality

$|a + b| \leq |a| + |b|$

is called the triangle inequality and

$|a + b| \leq \max (|a|, |b|)$

is called the strong triangle inequality.

10.2. Definition: A metric space (X, d) is said to be an ultrametric or a nonarchimedean metric space if for $x, y, z \in X$ we have

(*i*) $d(x, y) = 0$ iff $x = y$

(*ii*) $d(x, y) = d(y, x)$

(*iii*) $d(x, z) \leq \max(d(x, y), d(y, z))$

Remarks:

(*i*) A valued field (F|·|) is a metric space with the metric

$$d(a, b) = |a - b|$$

(*ii*) A nonarchimedean valued field is a nonarchimedean metric space or an ultrametric space with the ultrametric

$$d(a, b) = |a - b|$$

We shall now discuss some interesting properties of nonarchimedean-valued fields, which are not true for the Archimedean case.

Theorem 66: If (F, |.|) is nonarchimedean, then

$|a + b| = \max(|a|\ |b|)$ for $a \neq b$. To be more specific $|a + b| = |a|$ if $|a| > |b|$.

Proof. Let $|a| > |b|$.

Hence $\quad |a| + |a + b - b| \leq \max(|a + b|, |b|) \leq \max(|a + b|, |b|)$

$$= |a|$$

So, $\quad \max(|a + b|, |b|) = |a|$, *i.e.*, $|a + b| = |a|$

Theorem 67: All triangles in a nonarchimedean-valued field F are isosceles. In other words, if $a, b, c \in F$, then two of the three numbers

$$|a - b|, |b - c|, |c - a| \text{ are equal.}$$

Proof: Let $|c - a| \neq |b - c|$. Without any loss of generality, we assume that

$$|c - a| > |b - c|$$

We now show that $|a - b| = |c - a|$.

Now, $$|a - b| = |a - c + c - b| \leq \max(|c - a|, |b - c|)$$
$$= |c - a|$$

Further, $$|c - a| = |c - b + b - a| \leq \max(|b - c|, |a - b|)$$
$$= |a - b| \text{ since } |c - a| > |b - c|$$

Thus, $|a - b| = |c - a|$ and this completes the proof.

Since a nonarchimedean field F is a metric space, one can define Cauchy sequences and convergent sequences in F in the usual way. For example, a sequence $\{x_n\}$ in F is said to be Cauchy if for any $\in > 0$ there is some positive integer n_0 such that $|x_n - x_m| < \in$ for n, $m > n_0$. $\{x_n\}$ is said to be convergent to x if for every $\in > 0$, there exists n_0 such that

$$|x_n - x| < \in \text{ for } n > n_0$$

Also, we say that an infinite series in a nonarchimedean field F converges if the sequence of partial sums of the series converges.

As in the classical case, one can show that every convergent sequence is Cauchy in a nonarchimedean field F. If the converse holds, then the field is said to be complete.

Again, just as in the classical situation, one can show that if Σx_n converges then $x_n \to 0$. The converse is not true in the Archimedean (real or complex with usual absolute value) case. For example, if

$x_n = \dfrac{1}{n}$ then $x_n \to 0$ but Σx_n is not convergent.

It is interesting to note that the converse holds in the nonarchimedean case if the field F is complete and this is given in the following theorem.

Theorem 68: Let $x_n \to 0$. Then, Σx_n converges provided F is complete in the nonarchimedean valuation.

Proof: Let $x_n \to 0$. Then, for $m > n$,

$$\left|\sum_{i=n+1}^{n+m} x_i\right| \leq \max(|x_{n+1}|, \ldots, |x_{n+m}|)$$

$$\to 0 \quad \text{as} \quad n \to \infty$$

Hence, Σx_n converges.

Since a nonarchimedean-valued field F is metric space, we define open spheres and closed spheres in F in the usual way. The following interesting property holds in the nonarchimedean case.

Theorem 69: Let F be a nonarchimedean-valued field. Then, every point in the open sphere of x of radius r,

$S_r(x) = \{y \in F : |y - x| < r\}$ is a center, *i.e.*, if $y \in S_r(x)$ then $S_r(x) = S_r(y)$

Proof: Suppose $y \in S_r(x)$. Let $z \in S_r(y)$. We have to show that $|z - x| < r$. But $|y - x| < r$ and $|z - y| < r$. So, by strong triangle inequality, $|z - x| < r$. The reverse inclusion can be proved similarly.

Let X be a vector space over a valued field F. A mapping 11.11 : X $\to$ R is called a noun on X if for all $x, y \in X$ and $\lambda \in F$,

(*i*) $\| x \| \geq 0$, $\| x \| = 0$ iff $x = 0$

(*ii*) $\| \lambda x \| = | \lambda | \ \| x \|$,

(*iii*) $\| x + y \| \leq \| x \| + \| y \|$.

A vector space X with a noun defined on it is called a normed linear space (nls).

If $\| x + y \| \leq \max (\| x \|, \| y \|)$,

then the norm is called nonarchimedean.

We have

10.3. (*i*) All triangles in a nonarchimedean nls is isosceles,

(*ii*) If $x_n \to 0$, then Σx_n converges if X is a complete nonarchimedean nls,

(*iii*) In a nonarchimedean nls X, every point of

$S_r(x) = \{y \in X : \| y - x \| < r\}$ is a centre.

The proof is similar to theorems 67, 68, 69.

APPENDIX I

Set Theory

A set is a well-defined collection of certain distinguishable objects. The objects are called 'elements' or 'members' of the set.

We usually denote sets by capital letters such as A, B, C, ... and elements of a set by small letters such as *a*, *b*, *c*, ... To indicate that an element a is a member of a set A, we write

$$a \in A,$$

which is read '*a* is an element of A', or '*a* belongs to A' To indicate that the element a is not a member of the set A, we write

$$a \notin A,$$

which is read '*a* is not an element of A' or '*a* does not belong to A'. By a 'well-defined collection' is meant that given a set A and an object *a*, either $a \in A$, or $a \notin A$.

There are two standard notations available for designating a particular set. One way of describing a set is to list the elements of the set between braces. Thus the set of all positive integers that are less than 5 can be written as

$$\{1, 2, 3, 4\};$$

the set of all positive integers can be written as

$$\{1, 2, 3, ...\};$$

the set of all odd integers can be written as

$$\{\pm 1, \pm 3, \pm 5, ...\}$$

and the set of all the fourth roots of unity can be written as

$$\{1, i, -1, -i\}$$

With this notation an arbitrary set with six members may be written

$$\{a, b, c, d, e, f\}.$$

At times it is inconvenient or impossible to list all the elements of a set. In such cases, a set can often be described by specifying a property that the elements of the set have in common. For example, consider the set

$$A = \{1, 2, 3, 4, 5, 6\}$$

which has been described here by listing the elements. A is precisely the set of all positive integers that are less than 7 and this can be written as

$$A = \{x \mid x \text{ is a positive integer less than } 7\}$$

which is read A is the set of all x such that x is a positive integer less than 7. As another example consider the following expression

$$\{x \mid x \text{ is rational}\} \text{ or } \{x \mid x = \frac{m}{n}, m \text{ and } n \text{ are integers, } n \text{ is positive}\}$$

which is read 'The set of all x, such that x is rational', denotes the set of all real numbers which can be written as the quotient of two integers. The four sets of numbers described at the beginning of this paragraph can be written either way:

$$\{1, 2, 3, 4\} = \{x \mid x \text{ is positive integer less than } 5\}$$

$$\{1, 2, 3, ...\} = \{x \mid x \text{ is positive integer}\},$$

$$\{1, i, -1, -i\} = \{x \mid x \text{ is a complex number and } x^4 = 1\}$$

$$\{\pm 1, \pm 3, ...\} = \{x \mid x \text{ is an odd integer}\}$$

Two sets A and B are equal if and only if they have the same elements. This is known as the *axiom of extension*. In this case we write

$$A = B.$$

If A and B are not equal, then we write

$$A \neq B.$$

For example, if

$$A = \{1, 2, 3,\}$$

and $$B = \{x \mid x \text{ is a positive integer and } x^2 < 16\}$$

then $$A = B.$$

Sometimes one sees sets that have no members. For example, the sets

$$\{x \mid x \text{ is rational and } x^2 = 2\},$$

$$\{x \mid x \text{ is real and } x^2 = -1\},$$

have no members. A set with no members is called an *empty set* or a *null set* and such a set is denoted by the symbol ϕ. A set is said to be *finite* if it is empty or consists of n elements for some positive integer n; otherwise it is said to be *infinite.*

More precisely, the existence of the empty set follows from the axiom of specification which says that, 'To every set A and to every condition S(x) there corresponds a set B whose elements are exactly those elements x of A for which S(x) holds'. By applying this axiom to any set A with the condition '$x \neq x$' we get the empty set

$$\{x \in A \mid x \neq x\}.$$

If every element of a set A is also an element of set B, then we say that A is a subset of B or is contained in B and write

$$A \subseteq B.$$

We sometimes express this by saying that B is a superset of A (or contains A). $A \subseteq B$ allows for the possibility that A and B might be equal. If A is a subset of B and A is not equal to B, then we say that A is a proper subset of B or is properly contained in B and this is written as

$$A \subset B.$$

We can also express this by saying that B is a proper superset of A or properly contains A. The existence of subsets follows from the axiom of specification. It is customary to agree that empty set $\varnothing$ is a subset of every set.

It is often convenient to have a symbol for logical implication. If p and q are statements, then $p \Rightarrow q$, means that p implies q, or that, if p is true, then q is true. Similarly $p \Leftrightarrow q$ means that p is equivalent to q, or that p is true if and only if q is true.

If A is any set, then

$$A \subseteq A$$

Let A, B and C be three sets. Then

$$A \subseteq B \text{ and } B \subseteq A \Rightarrow A = B,$$

$$A \subseteq B \text{ and } B \subseteq C \Rightarrow A = C.$$

Note that

$$A = B \quad \Leftrightarrow \quad A \subseteq B \text{ and } B \subseteq A.$$

If a set A is not a subset of B we write

$$A \not\subset B$$

Algebra of Sets

The intersection of two sets A and B is the set of all elements that belong to both A and B, it is denoted by the symbol

$$A \cap B.$$

If $A \cap B \neq \phi$, we say that A intersects B. If $A \cap B = \phi$, then we say that A does not intersect B or that A and B are disjoint. That $A \cap B$ is a set follows from the axiom of specification if we write

$$A \cap B = \{x \in A \mid x \in B\} \text{ or } A \cap B = \{x \in B \mid x \in A\}.$$

The intersection of any collection of sets is the set of all elements that belong to every one of the sets in the collection.

The operation of forming intersections is commutative and associative:

$$A \cap B = B \cap A$$

and

$$A \cap (B \cap C) = (A \cap B) \cap C$$

It also has the property that

$$A \cap A = A, A \cap \phi = \phi.$$

The union of two sets A and B is the set of all elements that belong to either A or B or both, it is denoted by the symbol

$$A \cup B$$

The above definition can also be expressed as

$$A \cup B = \{x \mid x \in A \text{ or } x \in B\}$$

The operation of forming unions is also commutative and associative:

$$A \cup B = B \cup A$$

and $$A \cup (B \cup C) = (A \cup B) \cup C$$

It also has the properties that

$$A \cup A = A, \quad A \cup \phi = A$$

The union of any collection of sets is the set of all elements that belong to one or more of the sets in the collection. That it is a set follows from the axiom of union: For any collection of sets there is a set that contains all the elements that belong to at least one set of the given collection.

We have now defined two of the fundamental operations on sets and we shall now see how they are related to one another. The facts here are given by the distributive laws:

$$A \cap (B \cup C) = (A \cap B) \cup (A \cap C)$$

$$A \cup (B \cap C) = (A \cup B) \cap (A \cup C)$$

The next of our major operations on sets is the formation of complements.

Given any set A, it may be necessary to consider those elements that are not in A. Usually, however, there are elements not in A that are extraneous to the problem under consideration. For example, let A be the set of all positive integers. Clearly, if x is a negative integer, then $x \notin A$. However, if x is a cabbage then x is also not in A since cabbages are not positive integers. If we are concerned with a problem about integers then 'an element x not in A' is really meant 'an integer x that is not in A'. Cabbages are extraneous to the problem.

One way of eliminating extraneous elements is to specify in advance that all elements in the problem under consideration come from some fixed set U, called the *universal set*. For example, in a problem concerned with real numbers, we might agree to take the universal set U to be the set of real numbers and by 'an element x that is not in A' we must mean 'a real number x that is not in A'. In many problems the universal set is clear from the nature of the problem and is not explicitly stated.

If U is the universal set and A is a subset of U, then the set of all elements in U that are not members of A is called the complement of A; it is denoted by the symbol

$$A^C$$

The operation of forming complements has the following obvious properties:

$$(A^C)^C = A,\ U^C = \phi,\ \phi^C = U$$

$$A \cup A^C = U,\ A \cap A^C = \phi.$$

The following two important results, called *De Morgan's laws*, give relationship of complementation with union and intersection.

If A and B are any two sets, then

$$(A \cup B)^C = A^C \cap B^C$$

$$(A \cap B)^C = A^C \cup B^C.$$

We now prove the first De Morgan's law.

Let $x \in (A \cup B)^C$

$$\Leftrightarrow \quad x \notin A \cup B$$

$$\Leftrightarrow \quad x \notin A \text{ and } x \notin B$$

$$\Leftrightarrow \quad x \in A^C \text{ and } x \in B^C$$

$$\Leftrightarrow \quad x \in A^C \cap B^C.$$

Hence

$$(A \cup B)^C = A^C \cap B^C.$$

We now introduce the difference and symmetric difference of two sets. If A and B are two sets, we define

$$A - B = \{x \in A \mid x \notin B\}$$

That this is a set follows from the axiom of specification. It is easy to see that

$$A - B = A \cap B^C$$

We define the symmetric difference of two sets A and B as

$$A \,\Delta\, B = (A - B) \cup (B - A).$$

This operation is commutative and associative, *i.e.*,

$$A \,\Delta\, B = B \,\Delta\, A$$

$$A \,\Delta\, (B \,\Delta\, C) = (A \,\Delta\, B) \,\Delta\, C.$$

Also

$$A \,\Delta\, \phi = A, \; A \,\Delta\, A = \phi.$$

It can be checked easily that

$$A \,\Delta\, B = (A \cup B) - (A \cap B).$$

We have been considering the subsets of a given universal set U; do those subsets themselves constitute a set? The following axiom guarantees that the answer is yes.

Axiom of Powers: For each set there exists a collection of sets that contains among its elements all the subsets of the given set.

In other words, if U is a set, then there exists a set $\wp(U)$ such that if $X \subset U$, then $X \in \wp(U)$.

The set $\wp$ described above may be larger than wanted; it may contain elements other than the subsets of U. This is easy to remedy; just apply the axiom of specification to form the set $\{X \in \wp(U) : X \subset U\}$. Since for every X, a necessary and sufficient condition that X belongs to this set is that X be a subset of U, it follows that if we change notation and call this set $\wp(U)$ again, then

$$\wp(U) = \{X : X \subset U\}$$

The set $\wp(U)$ is called the power set of U; the axiom of extension guarantees its uniqueness.

Because the set $\wp(U)$ is very big in comparison with U, it is not easy to give examples. If $U = \phi$, then

$$\wp(U) = \{\phi\},\ \wp(\{a\}) = \{\phi, \{a\}\} \text{ and}$$

$$\wp(\{a, b\}) = \{\phi, \{a\}, \{b\}, \{a, b\}\}$$

The power set of triple has eight elements. The reader can probably guess (and is here by challenged to prove) the generalisation that includes all these statements: the power set of a finite set with n elements has 2^n elements.

RELATIONS

In many situations we are interested in paired data. An ordered pair (a, b) is a listing of two objects, a and b in a definite order. The element a is called the first entry and the element b is called the second entry of the ordered pair. Two ordered pairs are said to be equal if they list the same objects in the same order. Thus, for (a, b) and (c, d) to be equal, we must have

$$a = c \text{ and } b = d$$

Thus two ordered pairs (a, b) and (b, a), though involving the same elements, convey different information because of their differing order unless $a = b$.

We are now in a position to define the Cartesian product of sets. If A and B are two sets, the Cartesian product of A and B is the set of all ordered pairs (a, b) where $a \in A$ and $b \in B$, which is expressed by:

$$A \times B$$

Thus

$$A \times B = \{(a, b) \mid a \in A, b \in B\}$$

Example: If $A = \{a, b, c\}$

and $B = \{1, 2\}$

then $A \times B = \{(a, 1), (a, 2), (b, 1), (b, 2), (c, 1), (c, 2)\}$.

It is possible to define the Cartesian product of more than two sets. For example, if A_1, A_2, ... A_n, are n sets, where n is a positive integer, then the Cartesian product

$$A_1 \times A_2 \times ... \times A_n$$

is defined by

$$A_1 \times A_2 \times ... A_n = \{(a_1, ..., a_n) \mid a_1 \in A_1 ..., a_n \in A_n\}$$

A *relation* is any set of ordered pairs. Thus a relation is any set which is a subset of the Cartesian product of two sets. Let X and Y be any two sets (which may be equal, but need not be) and let R be a relation from X into Y. Then $R \subset X \times Y$. We define the domain of R to be the set, dom R = $\{x \in X : (x, y) \in R \text{ for some } y \in Y\}$ and the range of R to be the set, range R = $\{y \in Y : (x, y) \in R \text{ for some } x \in X\}$. The symbol R^{-1} denotes the inverse of R : $R^{-1} = \{\{y, x) : (x, y) \in R\}$.

There are two types of relations which often arise in mathematics: *equivalence relations* and *order relations*. We first discuss equivalence relations.

Let X be any non empty set. An equivalence relation on X is any relation $\sim \subset X \times X$ such that, for all x, y, z, in X we have

(*i*) $x \sim x$ (reflexive);

(*ii*) $x \sim y \Rightarrow y \sim x$ (symmetric);

(*iii*) $x \sim y$ and $y \sim z \Rightarrow x \sim z$ (transitive)

An equivalence relation on X has an intimate connection with decompositions of X into nonempty subsets which fill it out and have no elements in common with one another.

A partition of X is a disjoint class $\{X_i\}$ of nonempty subsets of X whose union is the full set X itself.

In other words, a partition of X is the result of splitting it, or subdividing it into nonempty subsets in such a way that each element of X belongs to one and only one of the given subsets.

We now show that each partition of X has associated with it a natural equivalence relation in X and, conversely, a given equivalence relation in X determines a natural partition of X.

Let X be a nonempty set. Assume that a partition of X is given. We now associate with this partition a relation '~' in X. We say that $x \sim y$ if x and y belong to the same partition set. It is easy to see that $\sim$ defines an equivalence relation in X.

Conversely, let $\sim$ be an equivalence relation in X. Let x be an arbitrary element of X. The set defined by $[x] = \{y : y \sim x\}$, which is the set of all elements of x which are equivalent to x, is called the equivalence set of x. We show that the class of all distinct equivalence sets forms a partition of X. By reflexivity $x \in [x]$ for each $x \in$ X, so each equivalence set is nonempty and their union is X. It remains to show that any two equivalence sets $[x_1]$ and $[x_2]$ are either disjoint or identical. Suppose that $[x_1]$ and $[x_2]$ are not disjoint. To be precise, let $z \in [x_1] \cap [x_2]$. Now $z \sim x_1$; and $z \sim x_2$, and by symmetry $x_1 \sim z$. Let $y \in [x_1]$, so that $y \sim x_1$; now, since $y \sim x_1$ and $x_1 \sim z$, transitivity shows that $y \sim z$. Since $y \sim z$ and $z \sim x_2$, again transitivity shows that $y \sim x_2$, so $y \in [x_2]$. Since y is an arbitrary element of $[x_1]$, this shows that

$$[x_1] \subseteq [x_2]$$

The same reasoning shows that $[x_2] \subseteq [x_1]$

Thus, $[x_1] = [x_2]$ and this completes the proof.

The following is a simple example of equivalence relation.

Example 2

Let Z be the set of all integers. For $a, b \in$ Z we define $a \sim b \Leftrightarrow a = b$.

Then it is evident that $\sim$ defines an equivalence relation in the set Z and each equivalence class consists of precisely one integer.

We now turn to the concept of *order relations.*

Let X be any nonempty set. A partial order relation (por) in X is a relation $\leq \subset$ X × X such that for all $x, y\ z \in$ X we have

(*i*) $x \leq x$ (reflexive);

(*ii*) $x \leq y$ and $y \leq x \Rightarrow x = y$ (antisymmetric);

(*iii*) $x \leq y$ and $y \leq z \Rightarrow x \leq z$ (transitive).

We sometimes write $x \leq y$ in the equivalent form $y \geq x$. A nonempty set X with a partial order relation defined on it is called a partially ordered set (poset). We now give three important and simple examples of partially ordered sets.

Example 3

Let X be the set of all positive integers, and $x \leq y$ mean that x divides y.

Example 4

Let X be the set of all real numbers, and $x \leq y$ mean that x is less than or equal to y in the usual sense.

Example 5

Let X be the family of all subsets of some universal set U and let $A \leq B$ mean $A \subseteq B$.

Two elements x and y in a poset are called *comparable* if either $x \leq y$ or $y \leq x$. Note that there may be pairs of elements in a poset which are not comparable. In Example 3, for example, the integers 3 and 7 are not comparable because neither divides the other; and in Example 5 if U has more than one element, there are always two subsets of U, neither of which is a subset of the other.

Some partial order relations possess a fourth property, in addition to the three required by the definition:

(*iv*) any two elements are comparable. A por with property (*iv*) is called a total (or linear) *order relation* and a set with a total (or linear) order relation defined on it is called a *totally ordered set* or a *linearly ordered set* or a chain.

FUNCTIONS

Let X and Y be any two sets (which may be equal but need not be). A function from X into Y is a rule 'f' which assigns to each element $x \in X$ a unique element $y \in Y$. The y which corresponds in this way to a given x is usually written $f(x)$, and is called the image of x under the rule 'f', or the value of 'f' at the element x. A function 'f' can also be viewed as a relation, that is, a subset of $X \times Y$ in which

$(x, y) \in f$ and $(x, z) \in f$ implies that $y = z$.

A function f from X into Y is often called a *mapping*, or a *map*, or a *transformation* from X, into Y. The set X is called the domain of f, written as 'dom f', and the set of all values of f for all x in X is called its range and is written as 'rng f'. A function whose range consists of just one element is called a constant function.

If f is a function from X into Y it is denoted by the symbol

$$f : X \to Y$$

If it is clear from the context what the sets X and Y are, or if there is no need of specifying them explicitly, it is common practice to identify the function $f : X \to Y$ by the rule f without mentioning the sets X and Y. If two definite sets X and Y are under discussion, then a mapping of X into Y is sometimes designated by

$$x \to y$$

where y is the image of x. For example, the function $f : R \to R$, where R denotes the set of all real numbers defined by

$$f(x) = x^2$$

can be designated by

$$x \to x^2.$$

A function f is called an *extension* of a function g (and g is called a *restriction* of f) if the domain of f contains the domain of g and $f(x) = g(x)$ for each x in the domain of g.

If $f : X \to Y$, let

$$G_f = \{(x, y) \in X \times Y : y = f(x)\}.$$

This set G_f is called the graph of f, the graph G_f is related to the function f in the sense that either one of them completely determines the other.

Consider a map $f : X \to Y$. Then clearly range of f is a subset of Y, that is range $f \subseteq Y$. If $f : X \to Y$ is such that range of f equals Y, then we call f surjective, or a mapping of X onto Y. If any two distinct elements in X have distinct images under f, then we call f

injective or a 1 – 1 mapping from X into Y. If f is both injective and surjective, then we call it bijective. If $f : X \to Y$ is bijective, then we can define its inverse mapping $f^{-1} : X \to Y$ as follows : for each $y \in Y$ there exists a unique element $x \in X$ such that $f(x) = y$ since f is 1 – 1 and onto; we then define x to be $f^{-1}(y)$.

Let X be a nonempty set. The identity mapping on X is the mapping of X onto itself which leaves each element of X fixed.

Suppose that $f : X \to Y$ and $g : Y \to Z$ be any two mappings. The product or composition of these two mappings, denoted by $gf : X \to Z$, is defined by

$$(gf)\ (x) = g(f(x))$$

We observe that the two mappings involved here are not arbitrary, for the set Y which contains the range of the first equals the domain of the second. More generally, the product of two mappings is meaningful whenever the range of the first is contained in the domain of the second.

Example 6

$f : R \to R$ be defined by

$$y = f(x) = x^2 + 1$$

and $g : R \to [-1, 1]$ be defined by

$$g(x) = \text{Sin } x.$$

Then $gf : R \to [-1, 1]$ is defined by

$$(gf)\ (x) = g\ (f(x)) = g(x^2 + 1)$$
$$= \text{Sin } (x^2 + 1).$$

Consider an arbitrary mapping $f : X \to Y$. If $A \subseteq X$, then its image $f(A)$ is the subset of Y defined by

$$f(A) = \{f(x) \mid x \in A\}$$

Similarly if $B \subseteq Y$, then its inverse image $f^{-1}(B)$ is the subset of X defined by

$$f^{-1}(B) = \{x \mid f(x) \in B\}.$$

The first mapping has the following properties:

$$f(\phi) = \phi; f(X) \subseteq Y;$$

$$A_1 \subseteq A_2 \Rightarrow f(A_1) \subseteq f(A_2)$$

$$f(\underset{\tau \in I}{\cup} A_i) = \underset{\tau \in I}{\cup} f(A_i)$$

$$f(\underset{\tau \in I}{\cap} A_i) \subseteq \underset{\tau \in I}{\cap} f(A_i)$$

The second one is much better behaved:

$$f^{-1}(\phi) = \phi; f^{-1}(Y) = X;$$

$$B_1 \subseteq B_2 \Rightarrow f^{-1}(B_1) \subseteq f^{-1}(B_2)$$

$$f^{-1}(\underset{\tau \in I}{\cup} B_i) = \underset{\tau \in I}{\cup} f^{-1}(B_i)$$

$$f^{-1}(\underset{\tau \in I}{\cap} B_i) = \underset{\tau \in I}{\cap} f^{-1}(B_{-i})$$

APPENDIX II

Groups and Rings

DEFINITIONS AND EXAMPLES

Definitions. Let G be a nonempty set. Let there be a mapping from G × G into G, called a binary composition, defined by

$$(a, b) \mapsto ab.$$

Then G along with this binary composition is called a *group* if it satisfies the following properties:

(*i*) For $a, b \in$ G we have $ab \in$ G (closure property),

(*ii*) For $a, b, c \in$ G we must have

$$a\,(bc) = (ab)\,c \text{ (associative law)}$$

(*iii*) There exists an element $e \in$ G called

the identity, such that for every $a \in$ G, $ae = ea = a$ (the existence of an identity),

(*iv*) For every $a \in$ G, there exists an element $a^{-1} \in$ G, called the inverse of a in G, such that $a\,a^{-1} = a^{-1}a = e$ (the existence of inverses).

The property (*i*) is a superfluous statement, it only tells again that the mapping $(a, b) \mapsto ab$ is form G × G into G, *i.e.*, it is a binary composition. G with property (*i*) only is called a *groupoid*, with (*i*) and (*ii*) is called a *semigroup* and with (*i*), (*ii*) and (*iii*) only is called a *monoid*.

A group G is called abelian or commutative if it has the following property.

(*v*) For $a, b \in$ G, $ab = ba$ (commutative law). Otherwise a group G is called *non-abelian*.

A group G is called finite if it is a *finite* set, otherwise it is called an *infinite group*. The number of elements of a group G is called the *order* of G and is denoted by o(G). This number is most interesting, useful and meaningful when G is a finite group.

Remark. In the above definition we have used multiplicative notation for the binary composition. If one uses additive notation the definition of a group reads as follows:

A nonempty set G along with a mapping $+ : G \times G \mapsto G$ defined by $(a, b) \mapsto a + b$ is called a group if

(*i*) $a, b, \in G$ implies $a + b \in G$,

(*ii*) $a, b, c \in G$ implies $a + (b + c) = (a + b) + c$,

(*iii*) For every $a \in G$, there exists an element $0 \in G$ such that $a + 0 = 0 + a = a$,

(*iv*) For every $a \in G$, there exists an element $(-a) \in G$ such that $a + (-a) = (-a) + a = 0$.

However, while proving the general results we shall use the multiplicative notation for the binary composition. Before we go further we shall first give some examples of groups leaving the verification for the reader as exercises.

Examples of Groups

(*a*) With respect to usual addition of numbers the following sets form groups:

(*i*) the singleton set consisting only of the number 0.

(*ii*) the set Z of all integers. The set of all nonnegative integers does not form a group.

(*iii*) the set of all even integers including 0. Observe that the set of all odd integers is not a group.

(*iv*) the set Q of all rational numbers.

(*v*) the set R of all real numbers.

(*vi*) the set C of all complex numbers.

(*vii*) the set of all complex numbers, whose real and imaginary parts are integers.

(*b*) The following sets of numbers form groups with respect to usual multiplication of numbers:

(*i*) the singleton set consisting only of the number 1.

(*ii*) the set $\{1, -1\}$.

(*iii*) $\{1, -1, i, -i\}$.

(*iv*) the set Z of all integers is not a group.

(*v*) the set of all positive rational numbers (excluding 0). The set of all nonzero rationals. Observe that the set of all negative rationals is not a group.

(*vi*) the set of all positive real numbers. Note that the negative real numbers do not form a group.

(*vii*) the set of all nonzero real numbers. The set R of all real numbers does not form a group.

(*viii*) the set of all nonzero complex numbers. Observe that the set C of all complex numbers does not form a group.

(*ix*) the unit circle $\{z \in C : |z| = 1\}$. This is called the circle group.

(*x*) the set $\{z \in C : z^n = 1\}$ where n is a fixed but arbitrary positive integer.

(*c*) The set R^n of all n-tuples of real numbers, *i.e.*, $R^n = \{x = (x_1, ..., x_n) : x_i \in R, i = 1, ..., n\}$, is a group with respect to coordinatewise addition defined by $x + y = (x_1 + y_1, ..., x_n + y_n)$ where $x = (x_1, ..., x_n)$ and $y = (y_1, ..., y_n)$. Similarly the set C^n of all n-tuples of complex number forms a group with respect to coordinate wise addition.

(*d*) The family of all subsets of a set U is a group with respect to the formation of symmetric difference of two sets A and B defined by $A \Delta B = (A - B) \cup (B - A)$.

It should be noted that if $U \neq \phi$, then the above family of sets does not constitute a group with respect to the formation of either unions or intersections.

(*e*) Let m be a positive integer and let Z_m be the set of all nonnegative integers less than m, *i.e.*, $Z_m = \{0, 1, ..., m - 1\}$. If a and b are in Z_m define their sum $a + b$ to be the remainder obtained when their usual sum is divided by m. Then Z_m is a group with respect to the operation defined above.

(*f*) The set of all 1 – 1 mappings of a nonempty set X onto itself is a group with respect to the multiplication (composition) of mappings defined as follows: if f and g are two such mappings and x is an arbitrary element of X, then $(fg)(x) = f(g(x))$.

(*g*) As a special case of the above example, let X = (1, 2, ..., n). This case is of particular importance. A 1 – 1 mapping of this set X onto itself is called a permutation and can be written in the convenient form

$$\begin{pmatrix} 1 & 2... & n \\ i_1 & i_2... & i_n \end{pmatrix}$$

where below each of the integers 1, 2, ..., n, is placed its image under the mapping.

For example, if $n = 4$ two such mappings p and q can be conveniently written in the form

$$p = \begin{pmatrix} 1 & 2 & 3 & 4 \\ 3 & 1 & 4 & 2 \end{pmatrix}, q = \begin{pmatrix} 1 & 2 & 3 & 4 \\ 4 & 1 & 3 & 2 \end{pmatrix}$$

We define the product pq of p and q by

$$pq = \begin{pmatrix} 1 & 2 & 3 & 4 \\ 2 & 3 & 4 & 1 \end{pmatrix}$$

The set of all permutations of n elements is denoted by S_n and this forms a group under the product of permutations. This group is called the symmetric group of degree n.

(*h*) Consider the set of all $m \times n$ (with m and n fixed positive integers) matrices with real or complex entries. If $A = (a_{jk})$ and

$B = (b_{jk})$ are two such matrices, then we define the sum A + B to be the matrix $C = (c_{jk})$ with elements

$$c_{jk} = a_{jk} + b_{jk},\ j = 1, ..., m,\ k = 1, ..., n.$$

The above set forms a group with respect to the operation of matrix addition defined as above. In particular the set of all square $n \times n$ matrices forms a group under matrix addition, but this set does not form a group under matrix multiplication. Any n-rowed square matrix A need not have an inverse A^{-1} such that

$$AA^{-1} = A^{-1}A = I,$$

where I is the n-rowed identity (unit) matrix. This holds if and only if A is nonsingular, that is, if and only if det. A $\neq 0$ where det. A denotes the determinant of A.

(*i*) The set of all nth roots of 1 is a group with respect to multiplication. If A denotes the set of all nth roots of 1, then A $= \{e^{2\pi ir/n} : r = 0, 1, 2, ..., (n-1)\}$.

If $a = e^{2\pi ir_1/n}, 0 \leq r_1 \leq n-1$ and $b = e^{2\pi ir_2/n}, 0 \leq r_2 \leq n-1$, then $ab = e^{2\pi i(r_1+r_2)/n}$. If $r_1 + r_2 \leq n-1$, then $ab \in$ A. If $r_1 + r_2 \geq n$, then assume that $r_1 + r_2 = tn + k$, $0 \leq k \leq n-1$ and then $ab = e^{2\pi i(tn+k)/n} = e^{2\pi it} . e^{2\pi ik/n} = e^{2\pi ik/n} \in$ A since $0 \leq k \leq n-1$.

The verification of other axioms of group are left to the reader. It should be noted that this group is abelian, for, $e^{2\pi ir_1/n} . e^{2\pi ir_2/n}$ $= e^{2\pi i(r_1+r_2)/n} = e^{2\pi i(r_2+r_1)/n} = e^{2\pi ir_2/n} . e^{2\pi ir_1/n}$.

Worked Out Examples

(*a*) Show that if every element of the group G is its own inverse, then G is abelian

Let $a, b \in$ G. Then $a^{-1}, b^{-1} \in$ G, $ab \in$ G and $(ab)^{-1} \in$ G. But $a = a^{-1}$, $b = b^{-1}$. Now $ab = (ab)^{-1} = b^{-1}a^{-1}, = ba$.

Hence G is abelian.

(*b*) If $a^2 = e$ for every element a in a group G, then show that G is abelian, where $a^n = a\ a\ ...\ a$ (n times).

We have $a^2 = aa = e$. But $aa^{-1} = e$.

Therefore $aa = e = aa^{-1}$ and so $a = a^{-1}$ for every $a \in G$. Now it follows from the previous example that G is abelian.

(*c*) If a group G has four elements, then show that it must be abelian.

Let $G = \{e, a, b, c\}$ where e is the unit element in G. If each element is its own inverse, then G is abelian. If not, assume that $a^{-1} = b$ and so $c^{-1} = c$. That is, $ab = ba = e$ and $cc = e$. Now

$$ac \neq e \text{ since } a^{-1} \neq c,$$

$$ac \neq a \text{ since } c \neq e,$$

$$ac \neq c \text{ since } a \neq e.$$

Therefore $ac = b$. Similarly it can be shown that $ca = b$.

Thus, $ac = ca$.

Again

$$bc \neq e \text{ since } b^{-1} \neq c,$$

$$bc \neq b \text{ since } c \neq e,$$

$$bc \neq c \text{ since } a \neq e.$$

Therefore, $bc = a$ and similarly, $cb = a$, so that $bc = cb$. Thus, G is abelian. The binary composition in G, which we call multiplication, is given by the following table:

	e	a	b	c
e	e	a	b	c
a	a	c	e	b
b	b	e	c	a
c	c	b	a	e

(*d*) If G is a group such that $(ab)^2 = a^2b^2$ for all $a, b \in G$, show that G must be abelian.

For all $a, b \in G$ we have

$$(ab)^2 = a^2b^2$$

ie, $$(ab)\,(ab) = (aa)\,(bb)$$

ie, $$a(ba)b = a(ab)b$$

ie, $$a(ba) = a(ab) \text{ (by right cancellation)}$$

ie, $ba = ab$ (by left cancellation)

Therefore G is abelian.

Example

(*a*) In any group G, $o(e) = 1$, where e denotes the identity element in G.

(*b*) In the multiplicative group $\{1, -1, i, -i\}$, $o(-1) = 2$, $o(i) = 4$.

(*c*) In the group $\{..., a^{-3}, a^{-2}, a^{-1}, a^0 = e, a, a^2, a^3, ...\}$, the order of any element is either 0 or ∞.

(*d*) If w denotes a cube root of 1, then in the group $\{1, w, w^2\}$, $0(1) = 1$, $0(w) = 0(w^2) = 3$.

Theorem

(*i*) Let G be a group and $a \in$ G. Then for any positive integer k, $o(a^k) \le o(a)$.

(*ii*) $o(a^{-1}) = o(a)$.

Proof

(*i*) Let $o(a) = n$ and $o(a^k) = m$. Therefore $a^n = e$ and $(a^k)^m = e$. So $(a^k)^n = (a^n)^k = e^k = e$. But m is the smallest positive integer for which $(a^k)^m = e$. So $m \le n$.

(*ii*) Let $o(a) = n$ and $o(a^{-1}) = m$. Now $(a^{-1})^n = a^{-n} = (a^n)^{-1} = e^{-1} = e$. Hence $m \le n$.

Similarly

$a^m = ((a^{-1})^{-1})^m = (a^{-1})^m = ((a^{-1})^m)^{-1} = e^{-1} = e$ and so $n \le m$.

Therefore $m = n$ and this completes the proof.

Theorem. The order of every element of a finite group G is finite.

Proof

Let $o(G) = n$ and let $a \in$ G be arbitrary. The elements $e, a, a^2, ..., a^n$ are $(n + 1)$ in number and $o(G) = n$, hence they are not all distinct. In other words, there exist r, s with $1 \le r \le s \le n$ such that $a^r = a^s$. Hence $e = a^r a^{-r} = a^s a^{-r} = a^{s-r} = a^k$ where $k = s - r$, $1 \le k \le n$.

Therefore $o(a)$ is finite and since a is arbitrary, this completes the proof.

Definition. Let G be a group. If for some $a \in$ G, each element of G is precisely of the form a^n for some $n \in$ Z, then G is said to be a *cyclic group* and a is said to be a *generator of* G.

Note that if G is a cyclic group and a is its generator, then

$$G = \{a^n : n \in Z\} = (a).$$

Examples

(*a*) Under addition

$$Z = \{..., -2, -1, 0, 1, 2, ...\}$$

is a cyclic group with 1 as the generator. Also under addition

$$\{..., -4, -2, 0, 2, 4, ...\}$$

is a cyclic group with 2 as generator.

(*b*) $\{..., a^{-2}, a^{-1}, a^0 = e, a', a^2, ...\}$ is a cyclic group with generator a.

(*c*) Under multiplication $\{1, -1, i, -i\}$ is a cyclic group with i as a generator, $-i$ being a generator also.

Theorem

(*i*) Every cyclic group is abelian

(*ii*) The order of a cyclic group is same as that of its generator.

Proof

(*i*) Let $G = (a) = \{e, a, a^2, a^3, ...\}$ be a cyclic group with a as the generator. Then for any $m, n \in Z$, $a^m, a^n \in G$ and $a^m a^n = a^{m+n} = a^{n+m} = a^n a^m$ and hence, G is abelian.

In the additive notation we write $G = \{1, a, 2a, 3a, ...\}$.

In this case $ma + na = (m + n)a = na + ma$ and thus, G is abelian.

(*ii*) Let G be a cyclic group with a generator $a \in$ G. Let $o(a) = n$, *i.e.*, $a^n = e$ and $a^m \neq e$ for $o < m < n$.

If $m > n$, then let $m = qn + r$, $o \leq r < n$.

So

$$a^m = a^{qn+r} = a^{qn} \,.\, a^r = (a^n)^q \,.\, a^r$$

$$= e^q \,.\, a^r = e \,.\, a^r = a^r.$$

Therefore, there are exactly n elements in the group and they are $\{e, a, a^e, ..., a^{n-1}\}$.

This completes the proof.

SUBGROUPS

Definitions. Let G be a group and let H be a nonempty subset of G. H is said to be a subgroup of G if, under the composition in G, H itself forms a group. If H is a subgroup of G, then we denote this by H < G.

Examples

(*a*) The singleton set $\{e\}$ and the whole group G., where e is the identity element of G, are subgroups of any group G. These are called trivial subgroups or improper subgroups. Any other subgroup is called a proper or nontrivial subgroup.

(*b*) The group of all even integers is a subgroup of the group of all integers under addition. But the set of all odd integers is not a subgroup of all integers under addition. Similarly the set of all positive integers is not a subgroup of all integers under addition.

(*c*) The group of all integers is a subgroup of all rational numbers under addition.

(*d*) The group of positive rational numbers under multiplication is a subgroup of the group of nonzero real numbers under multiplication.

(*e*) $\{1, -1, i, -i\}$ is a subgroup of the group of all nonzero complex numbers under multiplication.

(*f*) $\{1, -1\}$ is a subgroup of $\{1, -1, i, -i\}$ under multiplication.

Theorem. Let G be a group and let $\phi \neq H \subset G$.

Then H < G if and only if

(*i*) $a, b \in H$ implies that $ab \in H$,

(*ii*) $a \in$ H implies that $a^{-1} \in$ H.

Proof

If H < G, then it clearly follows from the definition that (*i*) and (*ii*) hold.

Conversely, suppose that $0 \neq \text{H} \subset \text{G}$ for which (*i*) and (*ii*) hold. Since associative law holds in G and $\text{H} \subset \text{G}$, it also holds in H. So in order to prove that H < G, it is now enough to show that $e \in$ H. For this, let $a \in$ H. By (*ii*) it follows that $a^{-1} \in$ H. Now by (*i*) we get $e = aa^{-1} \in$ H and this completes the proof.

Theorem. Let G be a group and let $\phi \neq \text{H} \subset \text{G}$. Then H < G if and only if $ab^{-1} \in$ H whenever $a, b \in$ H.

Proof

Let H < G and let $a, b \in$ H. Since $b \in$ H we have $b^{-1} \in$ H. Again since $a \in$ H, we have $ab^{-1} \in$ H.

Conversely, let $\phi \neq \text{H} \subset \text{G}$ for which $ab^{-1} \in$ H whenever $a, b \in$ H. Let $a \in$ H. Then $aa^{-1} = e \in$ H. Now for each $b \in$ H, we have $eb^{-1} = b^{-1} \in$ H. Let $a, b \in$ H. Then $b^{-1} \in$ H and hence $a(b^{-1})^{-1} = ab \in$ H. The associative law clearly holds and hence H < G.

Theorem. If H is a finite nonempty subset of a group G and H is closed under the composition, then H is a subgroup of G.

Proof

We need only to show that whenever $a \in$ H, then $a^{-1} \in$ H. Let $a \in$ H. Then since H is closed under the composition we have $aa = a^2 \in$ H, $a^2a = a^3 \in$ H, $a^3a = a^4 \in$ H, ..., $a^m \in$ H, ... Thus the infinite set of elements $a, a^2, a^3, ..., a^m$, ... must all lie in H. But since H is a finite set, there must be some repetitions in the above set of elements, *i.e.*, for some integers r and s with $r > s > o$ we must have $a^r = a^s$. Therefore $a^{r-s}a^{s-s} = e \in$ H. Since $r - s - 1 \geq 0$, $a^{r-s-1} \in$ H. Further since $aa^{r-s-1} = a^{r-s} = e$, we have $a^{-1} = a^{r-s-1}$. Thus $a^{-1} \in$ H. This completes the proof of the theorem.

DEFINITIONS AND EXAMPLES

Definitions. Let R be a nonempty set. Let there be two binary compositions, called addition and multiplication, defined on it. Then R is called a ring if it is

(*a*) an abelian group with respect to addition,

(*b*) a semigroup with respect to multiplication,

(*c*) the two distributive laws hold: for all $a, b, c \in R$,

$$a(b + c) = ab + ac$$

$$(b + c)a = ba + ca.$$

If we write explicitly then the definition of a ring will read as follows: A nonempty set R with two binary operations, addition and multiplication, defined on it is called a ring if for all $a, b, c \in R$,

(*i*) $a + b \in R$

(*ii*) $(a + b) + c = a + (b + c)$

(*iii*) There is an element $o \in R$ such that

$$a + o = o + a \text{ (for every } a \in R)$$

(*iv*) For every $a \in R$ there exists an element $-a$ in R such that

$$a + (-a) = (-a) + a = o$$

(*v*) $a + b = b + a$

(*vi*) $ab \in R$

(*vii*) $a(bc) = (ab)c$

(*viii*) $a(b + c) = ab + ac$

$$(b + c)a = ba + ca.$$

A ring R is said to be a *ring with unit element* (or a ring with identity if there is an element $1 \in R$, called the multiplicative identity, such that

$$a\,1 = 1\,a = a \text{ for every } a \in R.$$

R is called a *commutative ring* if the multiplication in R satisfies the commutative property, *i.e.*, if

$$ab = ba \text{ for all } a, b \in R.$$

A ring R is called *division ring* (or a *skew-field*) if all its nonzero elements form a group under multiplication.

A commutative division ring is called a *field*. In other words a ring R is called a field if all its nonzero elements form an abelian group under multiplication.

If R is a commutative ring, then $0 \neq a \in$ R is said to be a *zero-divisor* if there exists an element $0 \neq b \in$ R such that $ab = 0$.

A commutative ring is called and an *integral domain* if it has no zero-divisors.

Examples of Rings and Fields

(*a*) The following are rings with respect to usual addition and multiplication:

(*i*) The singleton set consisting only of the number 0

(*ii*) The set Z of all integers

(*iii*) The set of all even integers

(*iv*) The set Q of all rational numbers

(*v*) The set R of all real numbers

(*vi*) The set C of all complex numbers

It should be noted that the set in (*ii*) is an integral domain and the sets in (*iv*), (*v*) and (*vi*) are in fact fields.

(*b*) Any ring of subsets of a set U is a ring with respect to addition and multiplication defined by $A + B = A \Delta B$ and $AB = A \cap B$. The fact that a ring of sets is actually a ring is the reason for the name ring of sets.

(*c*) Let m be a positive integer and $Z_m = \{0, 1, ..., m - 1\}$. In Z_m define the sum $a + b$ and the product ab to be the remainders obtained when their usual sum and product are divided by m. Then Z_m is a ring with respect to the addition and multiplication defined as above. If m is any positive integer then Z_m is not always an integral domain. For example, if $m = 6$, then observe that $2.3 = 0$ and also $3.4 = 0$.

(*d*) The set $\{a+b\sqrt{2}: a, b \in Z\}$ is a ring under usual addition and multiplication of real numbers.

(*e*) Let

$$Z(i) = \{a + ib : a, b \in Z\}$$

Note that a complex number of the form $a + ib$ where $a, b \in Z$ is called a Gaussian integer and $Z(i)$ is called the set of all Gaussian integers. $Z(i)$ is a ring under the usual addition and multiplication of complex numbers. More precisely it is an Integral domain called the domain of Gaussian integers.

(*f*) The set $\{a+b\sqrt{2}: a, b \in Q\}$ is a field under usual addition and multiplication of real numbers.

(*g*) Let R denote the field of all real numbers. Then R × R is a field with respect to addition and multiplication defined as follows:

$$(a, b) + (c, d) = (a + c, b + d)$$

$$(a, b) . (c, d) = (ac - bd, ad + bc)$$

Note that (1, 0) and (0, 0) are the multiplicative and additive identities respectively. If $(a, b) \neq (o, o) \in R \times R$, then

$$(a, b)^{-1} = \left(\frac{a}{a^2+b^2}, \frac{-b}{a^2+b^2}\right)$$

since

$$(a,b)\left(\frac{a}{a^2+b^2}, \frac{-b}{a^2+b^2}\right) = \left(\frac{a^2+b^2}{a^2+b^2}, \frac{-ab+ab}{a^2+b^2}\right) = (1, 0).$$

(*h*) The set R × R is a ring under addition and multiplication defined as follows:

$$(a, b) + (c, d) = (a + c, b + d)$$

$$(a, b) . (c, d) = (ac, bd)$$

It is neither an integral domain nor a field. It is not an integral domain because,

$$(1, 0) . (0, 1) = (0, 0)$$

but $(1, 0) \neq (0, 0)$ and $(0, 1) \neq (0, 0)$.

(*i*) The set of all n-rowed square matrices forms a ring under addition and multiplication of matrices. This is not an integral domain. First observe that matrix multiplication is not commutative. For if $n = 2$, then

$$\begin{pmatrix}1 & 0\\0 & 0\end{pmatrix}\begin{pmatrix}0 & 1\\1 & 0\end{pmatrix}=\begin{pmatrix}0 & 1\\0 & 0\end{pmatrix} \text{but} \begin{pmatrix}0 & 1\\1 & 0\end{pmatrix}\begin{pmatrix}1 & 0\\0 & 0\end{pmatrix}=\begin{pmatrix}0 & 0\\1 & 0\end{pmatrix}.$$

Also by taking $n = 2$ we observe that

$$\begin{pmatrix}1 & 1\\2 & 2\end{pmatrix}\begin{pmatrix}-1 & 1\\1 & -1\end{pmatrix}=\begin{pmatrix}0 & 0\\0 & 0\end{pmatrix}.$$

This set is also not a skew-field. Because any non-zero square matrix A need not have a multiplicative inverse. This holds if and only if A is non-singular.

(*j*) Let M be the set of all 2 × 2 matrices of the form $\begin{pmatrix}\alpha & \beta\\-\overline{\beta} & \overline{\alpha}\end{pmatrix}$ where α, β are complex numbers and $\overline{\alpha}, \overline{\beta}$ are respectively the complex conjugates of α and β. Then M is a skew-field under addition and multiplication of matrices. If $A = \begin{pmatrix}\alpha & \beta\\-\overline{\beta} & \overline{\alpha}\end{pmatrix}$ is a nonzero element of M and if $D = \alpha\overline{\alpha}+\beta\overline{\beta}$, then the matrix $B = \dfrac{1}{D}\begin{pmatrix}\alpha & \beta\\-\overline{\beta} & \overline{\alpha}\end{pmatrix}$ is the inverse of A and belongs to M.

We now introduce a special class of rings which are important and useful in various applications.

Definition. A ring R is said to be a Boolean ring if $a^2 = a$ for every $a \in R$.

Examples

(*a*) The set {0, 1} under usual addition and multiplication.

(*b*) Any ring of sets under addition and multiplication defined by the symmetric difference and intersection respectively.

Worked Out Examples

(*a*) If R is a Boolean ring, then

(*i*) $a + a = 0$ for every $a \in R$.

(*ii*) $a + b = 0$ implies $a = b$.

For every $a \in$ R, $a + a \in$ R. Since R is a Boolean ring we have $(a + a) + 0 = (a + a) = (a + a)^2 = (a + a)(a + a)$ $= a(a + a) + a(a + a)$

$$= (a^2 + a^2) + (a^2 + a^2)$$

$$= (a + a)(a + a)$$

By cancellation law we have $a + a = 0$.

Let $a, b \in$ R be such that $a + b = 0$. But since R is a Boolean ring $a + a = 0$.

Therefore

$$a + a = 0 = a + b,$$

and by cancellation law if follows that $a = b$.

(*b*) Prove that every Boolean ring R is commutative.

Let R be a Boolean ring and let $a, b \in$ R. We have

$$\begin{aligned} a + b &= (a + b)^2 = (a + b)(a + b) \\ &= a(a + b) + b(a + b) \\ &= a^2 + ab + ba + b^2 \\ &= (a^2 + b^2) + ab + ba \\ &= (a + b) + (ab + ba) \end{aligned}$$

Now by cancellation law it follows that

$$ab + ba = 0$$

and hence by the previous example we have

$$ab = ba.$$

Thus, R is commutative.

Definition. Let R be a ring and S be a nonempty subset of R. S is said to be a subring of R if, under the compositions in R, S itself forms a ring. It is clear that a subring S of a ring R is a subgroup of the additive group R and that it is a semigroup under multiplication. Therefore we can give an alternative definition for a subring and this can be stated as a theorem as follows:

Theorem

Let R be a ring and let $\phi = S \subset R$. S is a subring of R if and only if

(*i*) $a, b \in S$ implies $a - b \in S$,

(*ii*) $a, b \in S$ implies $ab \in S$.

The proof of this theorem is easy and is left to the reader as an exercise.

Bibliography

1. Das, M.K. Number Theory (in Oriya language), Orissa Bureau of Text Book Preparation and Production (1977).
2. Chandra Sekharan, Introduction to Analytic Number theory.
3. Hardy and Wright, An Introduction to the Theory of Numbers.
4. Harry Pollard, The Theory of Algebraic Numbers.
5. Herstein, I.N. Topic in Algebra, Vikash Publishing House Pvt. Ltd., New Delhi (1985).
6. Nanda M.M., Modern Algebra (in oriya language), Orissa Bureau of Text Book Preparation and Production, Bhubaneswar (1980).
7. S. Nanda, Topics in Algebra, Allied publishers Pvt. Ltd, (2004).
8. Niven and Zuckerman, Introduction to the Theory of Numbers.
9. Tripathy N., Theory of Equations (in oriya language) Orissa Bureau of Text Book Preparation and Production, Bhubaneswar (1974).
10. Tom Apostal, Introduction to Analytic Number Theory.

Index